Youth Activism and Solidarity

From April 1986 until just after Nelson Mandela's release from prison in February 1990, supporters of the City of London Anti-Apartheid Group maintained a continuous protest, day and night, outside the South African Embassy in Central London. This book examines how and why a group of children, teenagers and young adults made themselves 'non-stop against apartheid', creating one of the most visible expressions of anti-apartheid solidarity in Britain.

Drawing on interviews with over 90 former participants in the Non-Stop Picket of the South African Embassy and extensive archival research using previously unstudied documents, this book offers new insights to the study of social movements and young people's lives. It theorises solidarity and the processes of adolescent development as social practices to provide a theoretically informed, argument-led analysis of how young activists build and practise solidarity.

Youth Activism and Solidarity: The Non-Stop Picket against Apartheid will be of interest to geographers, historians and a wide range of other social scientists concerned with the historical geography of the international anti-apartheid movement, social movement studies, contemporary British history and young people's activism and geopolitical agency.

Gavin Brown is an Associate Professor in the School of Geography, Geology and the Environment at the University of Leicester, UK. He is a cultural, historical and political geographer with an interest in protest movements, solidarity, and the geopolitics of sexual orientation and gender identity.

Helen Yaffe is a Lecturer in Economic and Social History at the University of Glasgow. She has focussed on Cuban economic history, political economy, Latin American development and the history of economics.

Routledge Spaces of Childhood and Youth Series
Edited by Peter Kraftl and John Horton

The *Routledge Spaces of Childhood and Youth Series* provides a forum for original, interdisciplinary and cutting-edge research to explore the lives of children and young people across the social sciences and humanities. Reflecting contemporary interest in spatial processes and metaphors across several disciplines, titles within the series explore a range of ways in which concepts such as space, place, spatiality, geographical scale, movement/mobilities, networks and flows may be deployed in childhood and youth scholarship. This series provides a forum for new theoretical, empirical and methodological perspectives and groundbreaking research that reflects the wealth of research currently being undertaken. Proposals that are cross disciplinary, comparative and/or use mixed or creative methods are particularly welcomed, as are proposals that offer critical perspectives on the role of spatial theory in understanding children and young people's lives. The series is aimed at upper-level undergraduates, research students and academics, appealing to geographers as well as the broader social sciences, arts and humanities.

Published:

Young Entrepreneurs in Sub-Saharan Africa
Edited by Katherine Gough and Thilde Langevang

Children, Young People and Care
Edited by John Horton and Michelle Pyer

Youth Activism and Solidarity: The Non-Stop Picket against Apartheid
Gavin Brown and Helen Yaffe

Youth Activism and Solidarity
The Non-Stop Picket against Apartheid

Gavin Brown and Helen Yaffe

LONDON AND NEW YORK

First published 2018 by Routledge

2 Park Square, Milton Park, Abingdon, Oxfordshire OX14 4RN
52 Vanderbilt Avenue, New York, NY 10017

Routledge is an imprint of the Taylor & Francis Group, an informa business

First issued in paperback 2019

Copyright © 2018 Gavin Brown and Helen Yaffe

The right of Gavin Brown and Helen Yaffe to be identified as authors of this work has been asserted by them in accordance with sections 77 and 78 of the Copyright, Designs and Patents Act 1988.

All rights reserved. No part of this book may be reprinted or reproduced or utilised in any form or by any electronic, mechanical, or other means, now known or hereafter invented, including photocopying and recording, or in any information storage or retrieval system, without permission in writing from the publishers.

Notice:
Product or corporate names may be trademarks or registered trademarks, and are used only for identification and explanation without intent to infringe.

British Library Cataloguing-in-Publication Data
A catalogue record for this book is available from the British Library

Library of Congress Cataloguing-in-Publication Data
A catalog record for this book has been requested.

ISBN: 978-1-138-82886-5 (hbk)
ISBN: 978-0-367-21895-9 (pbk)

Typeset in Times New Roman
by codeMantra

This book is dedicated to all those who fought against the injustice of apartheid in South Africa and internationally.

A number of key protagonists in our story – Norma, David and Steve Kitson – and former supporters of the Non-Stop Picket have died over the years. We remember them all, but would particularly like to pay tribute to five comrades who died while we were undertaking our research:

Solomon Odeleye (1955–2012)
Andrew Privett (1965–2012)
Zolile Hamilton Keke (1945–2013)
Ken Bodden (1950–2013)
Jacky Sutton (1965–2015)

Contents

List of figures ix
Acknowledgements xi
List of abbreviations xiii

1 South Africa and Britain in the 1980s 1

2 A non-stop protest in a non-stop world 20

3 Becoming non-stop 47

4 Being non-stop against apartheid 72

5 Defending the right to protest 103

6 Being unruly 126

7 Growing up through protest 147

8 'Until Mandela is free …' 173

9 Lessons and reflections 196

Bibliography 221
Index 239

List of figures

2.1	City Group leaflet advertising the launch of the Non-Stop Picket, 19 April 1986. Source: Gavin Brown.	42
3.1	*Fight Racism! Fight Imperialism!* banner in front of South African House, photographer unknown. Source: City Group.	49
3.2	City Group Singers perform on the Non-Stop Picket. Source: Photographer: Jon Kempster.	53
4.1	Arrest of protester during first anniversary rally, 19 April 1987, photographer unknown. Source: City Group.	91
4.2	Rally for the third anniversary of the Non-Stop Picket, April 1989. Source: Gavin Brown.	94
5.1	City Group postcard recording Councillor Bob Crossman, Mayor of Islington, lighting the brazier on the Non-Stop Picket, January 1987. Source: City Group.	108
5.2	Norma Kitson and David Yaffe defy the police ban on the Non-Stop Picket, 2 July 1987, photographer unknown. Source: City Group.	119
6.1	City Group poster produced for the *No Rights? No Flights!* campaign, 1988. Source: City Group.	136
7.1	Young picketers at the Surround the Embassy protest, 16 June 1988. Source: Photographer: Jon Kempster.	161
8.1	Picketers celebrate Mandela's release, 11 February 1990. Source: Photographer: Jon Kempster.	178

Acknowledgements

This book has had a long gestation. Gavin first thought about researching the history of the Non-Stop Picket in the late 1990s, but pursued other projects instead. Over the years, social media played a key role in helping him to re-establish contact with enough former picketers to make the research viable. This book primarily results from work undertaken for 'Non-Stop Against Apartheid: the spaces of transnational solidarity' funded by a research project grant [RPG-072] from the Leverhulme Trust. Helen joined the research project, bringing her skills as a historian and, importantly, negotiating access to the privately held archives of the City of London Anti-Apartheid Group [City Group].

We would like to thank the former participants in the Non-Stop Picket of the South African Embassy who shared their stories with us. It was a pleasure and an honour to be trusted with your memories. When City Group closed down in 1994, a group of members agreed to preserve the Group's papers with a view to eventually using them to contribute to the history of the group and its campaigns. We thank Carol Brickley and her comrades who helped look after the papers over the last two decades; and we thank the Revolutionary Communist Group for providing us with access to them. We still hope to find a way to transfer the papers to a secure public archive where they can be accessed by future researchers. We offer sincere thanks to Amandla Kitson for supporting this research and loaning us her brother's, Steven Kitson's, papers.

Our research project benefited from the advice and guidance of an advisory group who probably helped us more than they realise. Our appreciation, then, is due to Jenny Pickerill, Divya Tolia-Kelly, Jo Norcup, Kate Amis and three former City Group activists: Cat Wiener, Deirdre Healy and Mark Farmaner. We also thank Nicki for connecting us with other former non-stop picketers and giving us the benefit of her phenomenal memory. Jon Kempster, Rob Scott and Paul Mattsson, along with many others, shared photographs with us. Informal chats with scores of picketers on social media and in the bar at the British Film Institute have also been valuable.

Since 2011, we have given many talks about the history and geography of the Non-Stop Picket; the lessons of City Group's anti-apartheid campaigning;

and ways of theorising young people, geopolitics and practices of solidarity to a range of academic audiences and at various public engagement events. We thank all of those who invited us to talk, organised events for us and debated our ideas at London South Bank University, Trinity College Dublin, University College London and the universities of Birmingham, Leicester, Liverpool, Nottingham, and the West of England, as well as the Centre for Contemporary British History conference, International Conference of Historical Geographers and several Royal Geographical Society (with IBG) conferences. Thanks also to Ambrose Musiyiwa for inviting us to contribute to the Leicester Human Rights Arts and Film Festival.

Conversations, big and small, fraught and frivolous, with many academic colleagues have helped us shape our ideas. Amongst those whose interest, support and critical engagement has been particularly stimulating we would like to acknowledge Adam Barker, Margaret Byron, Ben Coles, Ruth Craggs, Andy Davies, Cesare di Feliciantonio, Klaus Dodds, Dave Featherstone, Mary Gilmartin, Gavin Grindon, Diarmaid Kelliher, Peter Kraftl, Maarten Loopmans, Kelvin Mason, Pete North, Kim Peters, Sasha Roseneil, Rob Skinner, Evan Smith, Simon Stevens and Alex Vasudevan. Working with Anna Feigenbaum, Fabian Frenzel and Patrick McCurdy to research protest camps helped Gavin think about the infrastructures and temporalities of the Non-Stop Picket in new ways. Gavin would also like to thank his PhD students Dave Ashby and Clara Rivas Alonso for their assistance at various stages of this research. The University of Leicester Geography Department provided Gavin with a semester's sabbatical in which to write the bulk of the first draft of this book, and a shorter leave of absence to complete the editing of it. He apologises to all his colleagues – academic and professional services – who have picked up additional work or dealt with his stress, as his mind was focussed on finishing this project. Countless Geography students have engaged enthusiastically with Gavin's teaching about apartheid and young people's opposition to it and helped him refine his ideas in the process.

We both thank Faye Leerink and Pris Corbett at Routledge for believing in this book and for their patience in waiting for its delivery.

Helen thanks all those who supported her throughout the research project, and especially her family, which grew in the duration.

As ever, Gavin could not have done it without the love, support, sustenance and stimulation of Joseph De Lappe.

List of abbreviations

AAM	Anti-Apartheid Movement
ANC	African National Congress
AWB	Afrikaner Weerstandsbeweging (Afrikaner Resistance Movement)
AZAPO	Azanian People's Organisation
BCM(A)	Black Consciousness Movement of Azania
CLAAG	City of London Anti-Apartheid Group[1]
COSATU	Congress of South African Trade Unions
CPGB	Communist Party of Great Britain
CPS	Crown Prosecution Service
LSE	London School of Economics
NSP	Non-Stop Picket
NUM	National Union of Miners
PAC	Pan Africanist Congress
RCG	Revolutionary Communist Group
SAA	South African Airways
SACP	South African Communist Party
SAEPC	South African Embassy Picket Campaign
SATIS	South Africa – the Imprisoned Society
SWANU	South West African National Union
SWAPO	South West African People's Organization
UDF	United Democratic Front
WRP	Workers' Revolutionary Party

Note

1 The acronym CLAAG was mostly used by the police and the AAM. The organisation's own supporters generally referred to themselves as City AA or City Group.

1 South Africa and Britain in the 1980s

In September 1986, the exiled South African activist Norma Kitson was interviewed on LBC radio in London. She opened the interview by stating,

> South Africa's my country and I want to return to it, as soon as I can. Yes! On the other hand, I mean, I am appalled at the increased brutality and repression in South Africa and… I'm *very urgent*, you know, to get maximum solidarity among British people, to help the liberation struggle end the apartheid regime.[1]

That interview was recorded five months into the Non-Stop Picket of the South African Embassy in London, which had been Norma Kitson's idea. This book tells the story of the Non-Stop Picket and the experiences and motivations of the young people from London and across the world who were inspired by her 'urgency' to build maximum anti-apartheid solidarity in Britain. This book is simultaneously a history of a particular moment in British anti-apartheid activism; a study in the spatiality of solidarity and contentious protest; and a study of the place of young people in those social movements and in the urban landscape of London in the 1980s.

The origins of apartheid

Apartheid was the system of racial segregation and discrimination that operated in South Africa in the late twentieth century. In Afrikaans, the term means 'separateness'. It was the main slogan of the Afrikaner National Party in their successful campaign in the South African (whites-only) general election of May 1948. Apartheid extended and institutionalised existing racial segregation, securing white minority rule in South Africa. Although the National Party's policy of apartheid extended the forms of racial segregation implemented under British colonial rule in South Africa over many decades, as an idea it took off in the late 1940s precisely because many white South Africans feared that the economic and military pressures of the Second World War had weakened segregation (Guelke 2005; Lester 1998).

Apartheid was implemented more systematically than previous forms of segregation in South Africa (De Jager et al. 2015). The Population Registration Act (1950) required the racial categorisation of the country's entire population (initially into 'native' [black], white, and coloured groups, although the naming and boundaries of these racial categories changed over time). A parallel law, the Group Areas Act (1950), served to enforce racial segregation in urban areas. The Immorality Act (1950) made sexual relations across the colour line illegal, and 'mixed marriages' were also prohibited. The Reservation of Separate Amenities Act (1953) removed any obligation from the government to provide equal provision of services to people in different racial groups. The practices of 'petty apartheid' meant that racially segregated railway carriages, park benches, beaches, public toilets, and other facilities were a pervasive feature of life in South African towns. Black and coloured people were stripped of any rights to representation in South Africa's national political institutions. In contrast, the Promotion of Bantu Self-Government Act (1959) implemented a system of ethnic 'homelands' based on the previous system of 'native reserves'. Everyone in the African population was allocated to one of eight ethnic population groups and expected to exercise political rights only in relation to the 'homeland' of each group. The homelands also served the racially segregated labour policy; African workers needed permission to work and reside in the majority of the national territory that was not allocated to the homelands. The repressive policing of these 'pass laws' was one of the most hated and contested aspects of apartheid.

By implementing apartheid, South Africa positioned itself in an ambivalent position geopolitically (Guelke 2005). On the one hand, the assertion of Afrikaner nationalism and the desire for independence from the British Empire can be interpreted in the context of broader nationalist tendencies at the time (albeit with a settler colonial character) (Lester 1998). However, intensified racial segregation and the unabashed white supremacism of many National Party supporters were at odds with broader anti-colonial trends internationally. The newly independent Indian government spoke out early on against the marginalisation of South Africa's Indian population. As the Cold War intensified, South Africa worked hard to position itself as a defender of white, Christian 'civilisation' against the worldwide 'communist threat'. Today, mainstream politicians across the political spectrum claim that they were against apartheid. This is disingenuous. As brutal and abhorrent as apartheid was, the ambivalence with which it was treated by many Western governments exposes some of the contradictions and hypocrisy of late twentieth-century liberal democracy. Despite their stated commitment to equality and human rights, many countries were prepared to do business with South Africa to benefit from the profits to be made from its highly unequal, racialized division of labour, where precarious African workers did the most dangerous work for the lowest wages.

Opposition to apartheid

Apartheid was opposed by people of all racial backgrounds in South Africa, including a minority of liberal and radical whites. However, to understand the history of opposition to apartheid, it is important to understand the history of different forms of African nationalism in South Africa. When the South African Native National Congress (SANNC), the precursor of the African National Congress (ANC), was formed in 1912, many of its leaders were members of the educated middle-class and hereditary elites who struggled for the inclusion of 'civilised men' like themselves within the existing racial hierarchy. In many ways, their nationalism was quite conservative and did not challenge assumptions about the superiority of European 'civilisation'. At the same time, there were other African nationalists in South Africa who were inspired by the more radical ideas of Marcus Garvey and WEB Du Bois, asserting the need to celebrate and reaffirm the value of African culture as an alternative to colonial rule (De Jager et al. 2015).

Frustrated by how moribund the ANC had become under its conservative leadership, in 1944 a group of young 'radicals' formed the ANC Youth League and started advocating for a programme of radical change, pursued through mass mobilizations on the streets. The Youth League included men like Nelson Mandela, Walter Sisulu, and Oliver Tambo who would go on to play a significant role in the ANC's opposition to apartheid over the next 50 years. It also included many Pan-Africanists such as Robert Mangaliso Sobukwe (Pogrund 1990) who (in 1958) would break away from the ANC to form the rival Pan Africanist Congress (PAC) in protest at the significant role of white and Indian communists in shaping the policies of the ANC.

As the new apartheid laws were implemented from 1950 onwards, the ANC implemented their Youth League's Plan of Action. Through a series of strikes, protests, and acts of civil disobedience, which deliberately broke apartheid's racist laws, the ANC's 1952 Defiance Campaign brought large numbers of ANC volunteers into joint action with militants from other racial groups for the first time. Although many of the leaders of the ANC Youth League (including Mandela and Sisulu) had previously held an anti-communist stance, the Defiance Campaign helped build trust between them and non-African communists.

Under pressure from the increasingly virulent anti-communism of the National Party government, and faced with being banned, the Communist Party of South Africa had voluntarily disbanded itself in 1950. Former party members sought to join other progressive organisations where they could have political influence – African communists joined the ANC, Indians joined the regional Indian congresses, and white communists formed a front organisation called the Congress of Democrats. In June 1955, the ANC and its allies held the Congress of the People in Kliptown, at which the multi-racial Congress Alliance adopted the Freedom Charter as its manifesto for a future non-racial South Africa. The first line of the

Freedom Charter proclaimed that South Africa belonged to all who lived in it, black or white. The politics of the Freedom Charter and the centrality of white militants in writing it finally provoked the Pan-Africanists to leave the ANC and form their own, African-led, organisation.

On 21 March 1960, the PAC organised a protest against the pass laws in Sharpeville Township in the Transvaal Province (now Gauteng). The police opened fire, killing 69 unarmed men and women, many shot in the back. The Sharpeville Massacre, as it became known, was a watershed in opposition to apartheid in South Africa and internationally. The South African government imposed a State of Emergency, detained thousands of militants and banned both the ANC and the PAC. In the context of this heightened state repression, both the ANC and PAC turned away from non-violent civil disobedience towards armed resistance to apartheid. Some members of both the ANC and the reconstituted South African Communist Party (SACP) had been advocating and secretly preparing for eventual armed struggle throughout the 1950s (as had militants from the Trotskyist Non-European Unity Movements and members of the Liberal Party who went on to form the Armed Resistance Movement).

Although it is often thought of as 'the armed wing of the ANC', Umkhonto we Sizwe (MK) was initially created in June 1961 as an organisation that was independent of the ANC but led by a High Command comprised of leading members of both the ANC and the SACP (Ellis 2013). Despite some successful sabotage operations against the infrastructure of the apartheid state (during 1962–1963), and no shortage of young volunteers wanting to be sent abroad for military training, the early years of MK were plagued by recklessness, complacency, and serious breaches of security (Suttner 2008: 35). Most of the members of the original High Command of MK were arrested at the Liliesleaf Farm near Rivonia in July 1963. Following these arrests, although some parts of the ANC's underground structures continued to operate, most of the organisation's leadership was either imprisoned or living in exile outside South Africa. The same was also true of the PAC (Kondlo 2009).

In the vacuum left by the decimation of the ANC and PAC inside South Africa, a new generation of anti-apartheid activists emerged in the late 1960s and early 1970s. Led by Steve Biko and Barney Pityana (amongst others), Black Consciousness was a new political philosophy that drew on Pan-Africanism, black theology, the ideas of Franz Fanon (Gibson 2011), and the Negritude philosophy of Aimé Césaire. Biko understood Black Consciousness as a simultaneous critique of the psychological impact white supremacy had on black people and a means to develop their capacity to imagine and prefiguratively self-organise alternative, decolonized ways of living.

Black Consciousness grew in influence throughout the early 1970s. It contributed to renewed trade union militancy in Durban in 1973 and inspired school students in Soweto to protest against the teaching of lessons in Afrikaans, which they regarded as 'the language of the oppressor' in

June 1976. When the police opened fire on the protesting school students, their revolt spread across the country. The Soweto Uprising demonstrated to the exiled leadership of the ANC that there was still the capacity for a mass anti-apartheid struggle inside South Africa. In the aftermath of the uprising, thousands of young people fled the country seeking military training. Because it had the most extensive external infrastructure, most were received by the ANC. The Soweto generation of militants would play a decisive role in shaping the struggles of the 1980s in South Africa.

The anti-apartheid movement in Britain (and internationally)

From the beginning, the apartheid system attracted condemnation from around the world. India and other post-colonial nations, as well as socialist bloc countries, were vociferous in their opposition to apartheid. As the ANC prepared for covert organising and potential armed struggle in the late 1950s and early 1960s, they sought financial, political, and military support from newly independent African nations, such as Ghana and Tanzania, but also sent their cadres to China, East Germany, and the Soviet Union for training (Ellis 2013; Shubin 2008; Suttner 2008). In the 1970s and 1980s, socialist Cuba sent some 350,000 troops to Angola to combat South Africa's invading army, which had occupied Namibia since 1915.

From the early 1960s onwards, but especially in the 1980s, there was a significant grassroots campaign against apartheid globally. Thörn (2006) suggests that the international movement against apartheid was the first genuinely global social movement to exist; and he has recorded evidence of anti-apartheid campaigning in about 100 different countries. As a result of those countries' colonial links with South Africa, the anti-apartheid movements in Britain and The Netherlands were among the most important in shaping the opposition internationally (although it is arguable that the US movement ultimately had the most significant impact in accelerating the end of apartheid, by pushing the US government to implement the Comprehensive Anti-Apartheid Act in 1986).

The formation of the Anti-Apartheid Movement (AAM) in Britain at the end of March 1960 was galvanised in reaction to the Sharpeville Massacre (Lodge 2011). Even so, the AAM did not appear out of nowhere. It developed from the Boycott Movement (initially the Boycott Committee) that had been formed the previous year, calling for the British public to boycott South African goods (Gurney 2000). From its inception, the Boycott Committee was a sub-committee of the Committee of African Organisations in London and comprised representatives of the Movement for Colonial Freedom, along with the ANC, the South African Indian Congress, and the Congress of Democrats, as well as individual campaigners motivated by a range of religious and political beliefs (Fieldhouse 2005; Skinner 2010). The AAM soon attracted the support of significant figures in the British Labour and Liberal parties but maintained an ambiguous relationship with the Communist Party, especially in the early 1960s.

The AAM was a national membership organisation, to which national bodies could also affiliate. Central to the movement was its national office in London, which also housed associated campaigns such as End Loans to South Africa (ELTSA), South Africa – the Imprisoned Society (SATIS) and the Namibia Support Committee. Five years after its formation, the AAM's membership had grown to around 2,500 people nationally. The membership stayed around this figure until the 1980s. By 1985–1986, the membership was around 7,500, and it peaked in March 1989 at 19,410 members (Fieldhouse 2005: 303). The membership was organised into local branches, which were uneven in their size and levels of activity.

The AAM's campaigning was remarkably consistent over the 35 years of its existence, setting the agenda for anti-apartheid campaigning in many other countries (Thörn 2006). It sought to bring international pressure to bear on the South African government, and to support and defend those South Africans (and Namibians) who opposed apartheid. A significant part of the AAM's work was encouraging individual and institutional consumers to boycott South African goods, culture, and sports. At the same time, it lobbied the British government to impose the economic sanctions and arms embargo against apartheid. The AAM also campaigned for the release of political prisoners in South Africa, opposing the execution of anti-apartheid prisoners, and organising practical support for prisoners and their families. However, it was not until 1980 that it launched a specific and prioritised campaign calling for the release of Nelson Mandela (Fieldhouse 2005).

Nationally (and locally), exiled supporters of the ANC and the SACP played an influential role in the leadership and conduct of the AAM's campaigns. Officially, the AAM offered support and solidarity to all South African liberation movements. However, in reality, the integration of ANC and SACP cadre into the leadership of the AAM ensured that it effectively recognised the ANC as 'the sole legitimate representative' of the South African people (Thomas 1996) and seldom campaigned directly for PAC or Black Consciousness affiliated prisoners, largely excluding their representatives from its public events (Williams 2015).

South Africa in the 1980s

The Non-Stop Picket came into being as a response to events in South Africa. It was launched in the context of the growth of a new mass anti-apartheid movement inside South Africa in the early 1980s and the government's repression of the township uprisings of 1984–1985 (Seekings 2000). Carol Brickley, the Convenor of the City of London Anti-Apartheid Group, which initiated the Non-Stop Picket, explained,

> There was an enormous build-up of militancy in the townships in South Africa from 1985 onwards which was extraordinary and very different from what had gone before. So that was the background to us

making any decisions. ... There was an enormous community uprising, effectively. In Alexandra Township they had workers councils, they were talking about making South Africa ungovernable at one point. So the Non-Stop Picket was a response to that, it wasn't that we suddenly decided that Nelson Mandela and all the political prisoners needed to be free. It was a way of bringing people's attention to what was going on in South Africa. Not only the prisoners but also the sheer brutality of the regime.[2]

In the mid-1980s the crisis of white rule in South Africa was coming to a head under increasing pressure from its opponents both inside and outside the country. The boycott of the new Tricameral Parliament by South Africa's Indian and coloured populations in August 1984, organised by the recently formed United Democratic Front (UDF), was a powerful blow to President Botha's attempts to reform apartheid. The use of the army to suppress the revolt in the country's townships, which began in September 1984, provoked both political action by the country's black trade unions and international condemnation (Guelke 2005). Around the world, anti-apartheid campaigners sought ways to increase their pressure on the South African government and its supporters. In a situation where diverse groups were challenging apartheid, the ANC sought to consolidate its position as the primary liberation movement (Thomas 1996).

According to the 1980 South African census, the majority of the South African population was under 21. It is not surprising then that youth played a central role in the politics of apartheid's final decade. Participation rates in secondary education were rapidly expanding, while failure rates in the matriculation exam were too. Faced with high levels of youth unemployment (anything up to 80 per cent for 18–26-year-olds in mid-1986 according to some estimates), "the experience of secondary education turned into a bottleneck of frustration rather than a window of opportunity" (van Kessel 2000: 51).[3] As with youthful engagements in the alter-globalization movements of the early 2000s (Juris and Pleyers 2009) and more recent 'no borders' organising (Burridge 2010), young people in the 1980s used "their common identity as young people or as a 'generation' as a starting point for critique and action" (Jeffrey 2013: 148) against apartheid across national borders. As many of the young participants in the Non-Stop Picket explained to us (see Chapter 3), although they often already held anti-racist beliefs, the repeated impact of seeing the repression of black youths in South Africa on the television news often provoked them into taking action against apartheid.

Birth of the UDF

The UDF was a wide-ranging alliance of community-based organisations inside South Africa that came together in August 1983 to oppose and challenge the limited reforms to the South African Constitution, which had been enacted that year (Seekings 2000). The Republic of South Africa Constitution Act of 1983 set up a Tricameral Parliament with each racially segregated chamber having jurisdiction over its ethnic constituency's 'own affairs'. This new constitution continued to legally enshrine the racial segregation of South Africa's population, perpetuated the disenfranchisement of the black African majority, and sustained the disproportionate political and economic power of the white minority population (Guelke 2005; Thörn 2006). In an attempt to give credibility to the new system, the government permitted South Africa's non-white populations a higher degree of freedom to express subordinate political views. In cohering the UDF as a mass democratic movement, opponents of apartheid took full advantage of this new-found freedom. While the new constitution was intended to dissipate anti-apartheid sentiment by the partial incorporation of the political aspirations of Indian and coloured communities, it had the opposite effect – it drew attention to the continuing disenfranchisement of the African population and intensified their resentment. The UDF represented the first significant revival of ANC-aligned organisations inside South Africa since the 1960s, when repression, imprisonment, and exile had decimated their organisational structures. The UDF provided a (semi-legal) infrastructure through which the ANC could rebuild its underground operations inside the country (Suttner 2008).

One of the key strengths of the UDF was its breadth and its ability to weave together the various strands of resistant practices that had been developed by sectors of society over the previous decade. Local civic groups from the townships, student groups, youth groups, women's groups, trade unions, and religious organisations, amongst others, were present at the UDF's founding conference (Lester 1998: 201). There were a variety of different ideological tendencies within the UDF, ranging from African nationalists committed to the policies advocated in the Freedom Charter, to socialists who understood apartheid as a specific form of capitalism and saw black workers as the revolutionary vanguard (Lester 1998). Alongside these tendencies, van Kessel (2000) recognises that many participants were also motivated by a belief in forms of Christianity inspired by liberation theology. Through the civics and unions, the UDF wove together local grievances about rents, evictions, and transport fares, with opposition to apartheid and state repression.

Many UDF publications in the mid-1980s proclaimed a vanguard role in the liberation movement for a 'student-worker alliance', and the 'stayaways' (generalised mass strikes) of November 1984, which protested the use of troops to repress trade union struggles, in particular were celebrated as an indication of the potential for such an alliance (van Kessel 2000: 67). The importance of this alliance was echoed by many of the youth organisations aligned with the UDF, and the ANC themselves promoted the role of the 'young lions' (politicised young 'warriors') in engaging the forces of apartheid on South Africa's streets. However, what the leading role of youth meant in practice, and who was considered to be a youth in this context, varied significantly. The more educated leaderships of youth and student organisations drew the political lesson from the school students' uprising in June 1976: that the youth could only succeed by accepting the leading role of the working-class in the liberation struggle. And yet, "many less sophisticated youth had no patience with theoretical and strategizing sessions: they wanted action" (van Kessel 2000: 67). As the struggle broadened, with the renewed township revolts of the mid-1980s, it was youthful street fighters who were challenging the apartheid state on the streets of their townships.

Van Kessel (2000: 67–68) argues that the undifferentiated use of the term 'youth' in relation to the anti-apartheid struggle in the 1980s is unhelpful and should not be equated with all young people (not even all 'non-white' young people) in South Africa at the time. Youth, in this context, is very closely linked to political activity. With on-going and significant disruption to schooling and high levels of unemployment amongst younger people, "*youth* could be a long-term career" (van Kessel 2000: 68). At times, 'youth' could include any high school graduate who was politically engaged and had not yet started a family. What youth meant and how it was understood in 1980s South Africa, like so much else, was highly differentiated by race and class. There was a huge growth in political youth organisations in the first half of the 1980s. Their leadership tended to be formed of school and university students as well as (relatively) well-educated young workers – as van Kessel (2000: 68) acknowledges, "they were the successes rather than the failures of the school system". As violent confrontations with the police and army became more commonplace, a less well-educated leadership emerged on the streets. Their ranks were highly gendered, with little place tolerated for young women in the leadership of this more violent, macho youth politics (Suttner 2008).

Throughout the early 1980s, the UDF and its affiliates significantly increased the pressure on the apartheid state, questioning its legitimacy and challenging its power. In addition to boycotting the Tricameral Parliament, the UDF organised boycotts of elections to the new 'black local authorities', starving them of revenue through rent boycotts in many townships. With barely 20 per cent of eligible voters participating in the 1983 local elections to these councils, the civics (working class, community-based civil society organisations) extended their operations to set up parallel systems of local governance, including (in some cases) the establishment of informal people's courts. In September 1984, protests against a rent increase in the five townships of the Vaal Triangle, including Boipatong and Sharpeville, turned violent and erupted into a full-scale revolt. Resentment about rent increases was amplified in a situation where local councillors in the widely boycotted councils were seen as collaborators with apartheid. The Federation of South African Trade Unions (FOSATU) had played an important role in organising rent boycotts in the Vaal Triangle; but the state's repressive reaction to the uprising further politicised the trade unions, which led to a two-day national strike in November 1984 (Bell 2009). The uprising and political violence spread and continued well into the following year when a State of Emergency was declared in July 1985 in 36 districts in an attempt to regain control of pockets of the country that had become increasingly 'ungovernable' (Thörn 2006). Throughout the insurrection, slogans, songs, and emblems of the ANC became increasingly visible and audible across South Africa, despite the organisation having been banned since 1960 (Lester 1998: 2005). Although the ANC had improved their capacity to infiltrate guerrillas back into the townships after training in Angola, Mozambique, and elsewhere, the township revolt of 1984–1985 gave an exaggerated impression of their actual numbers and gave the ANC renewed political legitimacy, nationally and internationally, as a leading force in the liberation struggle (Lester 1998; Suttner 2008).

By the summer of 1985, Botha and the South African government were under increasing pressure from both inside the country and internationally (Guelke 2005). The township revolt inspired increased anti-apartheid campaigning in many countries, including daily protests for a year outside the South African consulate in Washington, DC (Metz 1986).[4] On 31 July 1985, under pressure from the international anti-apartheid movement, the Chase Manhattan Bank declared that it would not be extending any further loans to South Africa. Botha's much-anticipated 'Rubicon speech' on 15 August offered only entrenchment rather than either significant reforms or an announcement of the end of apartheid (that many had expected). Faced with this intransigence, and with a growing mass movement against apartheid inside South Africa and internationally, more international banks refused to renew loans to South Africa, the pressure for international economic sanctions intensified, and the value of the Rand fell by a third in a week. Just as the Sharpeville Massacre in 1960 and the Soweto uprising of 1976 had inspired international opposition to apartheid, so too did the daily news reports of township youths confronting the brutal repression of the South African police and military in 1984–1985. In the mid-1980s, international anti-apartheid solidarity campaigns grew significantly and exerted additional pressure (directly and indirectly) on the South African authorities, mirroring the pressure of the mass movement on the country's streets (Fieldhouse 2005; Thörn 2006).

London in the mid-1980s: becoming a 'global' city

This introductory chapter locates the Non-Stop Picket within the changing social and political geographies that connected London to the rest of the world in the mid-1980s. In the midst of Margaret Thatcher's government, London was changing fast. Indeed, the outcome of political and economic struggles within London (and between London and other regions of the country) was central to the assertion of a neoliberal hegemony in the UK (Massey 2007: 80). The launch of the Non-Stop Picket on 19 April 1986 came less than three weeks after the Thatcher government had finally abolished the Greater London Council (GLC), leaving the capital with no directly elected, city-wide regional government for the next 14 years (Massey 2007: 17). The taken-for-granted social values of the post-Second World War welfare state were increasingly under attack, and new 'aspirational' values were coming to the fore (McSmith 2011). The pace of life in London was accelerating; the capital was reasserting itself as a connected, global city, attracting young migrants and visitors from across the world. If the pace of change was often exciting, it was also unsettling and discomforting: youth unemployment was high, and the young homeless were a visible presence on the streets of the West End. The Non-Stop Picket offered excitement and a sense of purpose to diverse young people who had not (yet) found their place in a changing city.

In the early 1980s, London was a city that felt like it was in decline. However, working-class London had seldom been associated with the types of place-specific 'militant particularisms' that were associated with mining communities and other groups of highly unionised workers, such as those at Oxford's Cowley car plant (Featherstone 2008; Harvey and Williams 1995). The collapse of the London docks and manufacturing industry during the preceding decade had blighted the city. Inner-city decline was an issue that concerned politicians and citizens alike. Doreen Massey (2007: 31) described this moment as "the final crumbling of an imperial geography". The spatial division of labour that had been established at the height of the British Empire was being remade and so was London's place in the world.

If the decline of the imperial trading networks and production base brought poverty and unemployment to many working-class Londoners, then the financial side of the city's imperial role was about to reinvent itself and change London's relationship to other parts of the world.

In the 1980s, the possibilities for London's future were still being fought out. Just as the miners fought (and lost) their battle to defend their industry and their working conditions, so too print workers in London fought the implementation of new working conditions and technologies that they understood as a threat to the privileged working conditions achieved by the strength of their trade union organising (Richardson 2003). In the east of London, local communities fought to shape the jobs, services, and land use that would replace the docks (Butler and Rustin 1996).

Between 1939 and 1981, London's population shrank by 1.8 million people (roughly 20 per cent of its pre-Second World War peak). By the end of the 1980s, as London's economy was realigned, the city's population began to grow again (Massey 2007). Between 1978 and 2000, employment in finance and banking services increased by 81 per cent, while manufacturing employment shrunk by 63 per cent (and, even then, there is evidence to suggest that many of the 'manufacturing' jobs that remained were administrative and managerial jobs in the headquarters of manufacturing companies that actually made goods elsewhere in the world) (Massey 2007: 32). As the world economy shifted towards greater reliance on financial capitalism and governments such as Thatcher's asserted the free-market ideology we now know as neoliberalism, there was a spatial reorganisation of the global economy. Financial services and the associated advanced producer services became concentrated in a small number of 'global cities' that served as the 'command and control centres' of the new economy (Sassen 1991). The 'Big Bang' deregulation of financial services in the UK, which took place in October 1986, opened up the City of London to foreign banks and financial services firms, but it also changed the place of financial services within the UK economy at a time when public utilities were being privatised and individual citizens were being encouraged to take greater financial responsibility for their own social reproduction (Buck et al. 2002).

Geographers have spent years debating and contesting the normative power of measuring the 'global-ness' of cities in this way, and prioritizing urban geographies based on concentrations of financial services (Amin and Graham 1997; Robinson 2006). Our purpose here is less to engage in those debates than to note that, at the time the Non-Stop Picket took place, the economic geography within London was changing, as was the city's place within global geoeconomic and geopolitical networks. Nevertheless, London's capacity to reposition itself so effectively at this time was itself a product of the city's central role as a colonial and imperialist metropole, as well as the political, economic, and social infrastructures that this had put in place over the preceding centuries (King 1990; Jacobs 1996).

As an imperial metropole, the social and political elites from across the (former) Empire were educated in London, including those who would lead their countries to post-colonial independence (Bunce and Field 2014; Høgsbjerg 2014; Matera 2015; Smith 2016). In the post-Second World War period, workers were recruited from across the Commonwealth to fill labour shortages in the lower-paid echelons of the public sector, in transport, and in manufacturing. In the 1980s, London still served as a global hub in which anti-colonial exiles from across the world converged and networked with each other (Centre for Contemporary Cultural Studies 1982; Virdee 2014a, 2014b). At the same time, the children of the post-war migrants were playing an increasingly visible role in London's social, cultural, and political life. On 28 September 1985, not much more than six months before the launch of the Picket, rioting had broken out in Brixton (at the time, a poor area of London with a large black population), for the second time in four years, in response to the shooting of Cherry Groce by Metropolitan Police officers, as they raided her home, looking for her son Michael, who was wanted for firearms offenses (Benyon and Solomos 1988). The riots spread to other multicultural areas of inner London, including Peckham and Tottenham, fuelled by high levels of youth unemployment and anger at police racism.

London's new status as a 'global city' also attracted new migrants to the city and produced a new migrant division of labour (Wills et al. 2010). The repositioning of London as a global city fostered new international links beyond the capital, but it also shaped new experiences of cosmopolitan conviviality amongst the city's (younger) residents (Gilroy 2004; Keith 2005).

Youth in the 1980s

In April 1986, when the Non-Stop Picket began, 3.3 million people were unemployed in Britain. Of this total, there were 1.25 million 16–24-year-olds who were unemployed, including a quarter of all 18-year-olds nationally (Abercrombie et al. 1988). At the time, less than 15 per cent of 18-year-olds went to university. In other words, despite high youth unemployment, the majority of people in their late teens were able to find employment; albeit relatively unskilled, low-waged work.

Despite the Thatcher government's attacks on many aspects of the welfare state, young people could still access unemployment and housing benefits when they were not in work. Even young people on low incomes could generally afford to rent accommodation in inner London and live independently of their parents from a young age. Although the first Thatcher government had begun the process of selling off council housing to sitting tenants (Evans 2013), young people could still access cheap social housing (especially in those estates and properties that were harder to let to families). For those young people unable (or unwilling) to rent accommodation in either the public or private rental sectors, squatting was common in inner London and supported by informal infrastructures of its own (McFarlane and Vasudevan 2014; Vasudevan 2017). Even so, a visible minority of unemployed young people, without the social and cultural capital to access these different forms of housing, found themselves homeless and sleeping on the streets of Central London (Oldman 2004; Turner 2010).

British politics and international geopolitics in the 1980s made the period unsettling and frightening for many young people. The combination of high unemployment, rapid social change, the apparent threat of nuclear war (especially in the early to mid-1980s), and the panic of the early years of the AIDS crisis made many young people fear for the(ir) future. For a minority of young people, like those who joined the Non-Stop Picket, this uncertainty inspired them to engage in campaigning and protest in the hope of shaping the world they would grow up in. But many more young people (including those who were politically active) responded to the state of the world despondently and, at times, nihilistically. At the same time, many young people were also open to the materialistic, aspirational promises of Thatcherism (Evans 2013; Jackson and Saunders 2012; Pilcher and Wragg 1996); and it is unlikely that those young people who subjectively understood themselves as opposed to Thatcherism were immune from these attitudes and influences. In many ways, young people in 1980s' Britain displayed all the contradictions of a society in rapid flux. In their attitudes to gender, sexuality, and race, young people were often more liberal than the rest of society. For example, throughout the 1980s, the percentage of British adults expressing the belief that homosexuality was wrong actually increased; but, by the end of the decade, half of people aged under 25 expressed their acceptance of homosexuality[5] (Weeks 2007: 17–18).

In thinking about young people and politics, there is one obvious point that needs to be made: there are asymmetries of power between adults and young people. Adults use power structures of various kinds to define and limit young people's life worlds and spatial practices (Skelton 2010). Young people's lives are affected by political decisions made by adults over which they seldom have any formal influence. And yet, of course, young people also wield political power too and sometimes can use their relatively disempowered position to tactical political advantage (Bosco 2010).

Peter Hopkins (2010: 136) notes that political geographers have largely overlooked young people from their work, citing Philo and Smith (2003: 103) in this regard:

> The sub-discipline of political geography has never shown any special interest in children and young people, for the understandable reason that people below voting age cannot and do not have much direct influence on the obviously 'political' phenomena and structures (to do with nations, states, federations, elections, geopolitics, boundaries and the like) that have long been the staple subject-matter of political geography.

Hopkins (2010) suggests that it is important to consider that young people may think about and engage with politics in ways that are different from those conventionally practised by adults (recognising that these differences are shaped as much by global inequalities as by generational effects). Frequently, debates about children and young people's political practice conclude that they "do politics in complex, innovative, melded and liminal ways and in a variety of spaces" (Skelton 2013a: R4; see also Mitchell and Elwood 2012; Skelton 2010; Wood 2012). For Kallio and Häkli (2011), young people carve out a political space for themselves at the meeting points between children's and adults' social worlds. Processes of socialisation occur in these spaces and bind young people to the worlds of *both* children and adults (Wood 2012: 338). There is a clash here between approaches that understand children as "citizens-in-the-making who develop into actual political actors and engaged citizens only when they reach adulthood" (Gordon 2008: 31; Skelton 2010), rather than political actors in their own right. This requires reframing politics as something other than just an adult pursuit without reducing young people's political engagements to the status of a 'hobby' or a 'lifestyle choice'.

This book, then, contributes both to the expanding literature on the political geography of young people's lives and their geopolitical agency (Benwell and Hopkins 2016; Hörschelmann 2008; Pain et al. 2010); as well as developing the still scarce attention to children and young people's geographies in the (recent) past (Dickens and MacDonald 2015; Mills 2011, 2012, 2015). By examining the role of children and young people in British anti-apartheid solidarity, we offer a contemporary historical approach to young people's geographies. By positioning our analysis of these young people's activism within the context of the broader geographies of young people's lives in 1980s' Britain, we look beyond easy tropes of adolescence as a time of experimentation and identity flux to consider how growing up in Thatcher's Britain profoundly changed how young people could think about their futures and their social and personal identities. Giddens (1991) famously noted a generational shift from an 'emancipatory politics', which fought for liberation and the improvement of collective life chances, towards a more highly individualised 'life politics' focussed on the "politics of self-actualisation in a reflexively ordered environment" (Giddens 1991: 214). Under these conditions, we consider how practices of political solidarity might have created other ways to 'transition' to adulthood. City Group (short for City of London Anti-Apartheid Group) and the Non-Stop Picket still articulated an emancipatory politics, but it could not totally escape the new 'life politics' of the time either.

Scope of the study

This book draws on interviews with former participants in the Non-Stop Picket and a range of archival material from that time. Prior to starting this research project, Gavin expected to interview somewhere in the region of 30 people. In the end, we interviewed 85 people who had been regular participants in the Non-Stop Picket. They were involved for varying lengths of time and with different levels of intensity and commitment. We also interviewed eight people who were close supporters of the picket – not necessarily people who spent a lot of time there, but high-profile politicians and public figures who attended periodically and could be relied on for vocal support at key times. They include some of the solicitors who helped defend arrested picketers in court.

We attempted to obtain a range of information from the Metropolitan Police about their policing of the Non-Stop Picket. They did, eventually, release some material – mostly their file copies of City Group publications, but most of our enquiries were met with standard 'neither confirm, nor deny' responses. Consequently, although it had not been part of our initial plan, a letter to the magazine of the National Association of Retired Police Officers helped us to track down and interview eight retired officers, of various ranks, who had been involved in policing City Group's protests in the mid-1980s.

As our participants are now dispersed around the world, these interviews were conducted through various modes. More than half of the interviews were conducted face-to-face; about a quarter were conducted online via Skype; a couple over the phone; and the remaining participants answered our standard questions in the form of an extended, qualitative questionnaire shared via email. In this way, we were able to record the stories of former picketers spanning an age range of nearly 50 years, and who were living in 11 countries (from Brazil to South Africa) in addition to Britain.

When City Group ceased to operate at the end of apartheid, some of the remaining members of the Group made plans to preserve the Group's archive with a view to publishing their story. That publication never happened, but we benefited from the decision to preserve a historic record of their anti-apartheid campaigning. For nearly 20 years, all of City Group's accumulated paperwork from their office – an archive spanning 12 years of activity (1982–94) – had been stored between Carol Brickley's workplace and the offices of the Revolutionary Communist Group. We were lucky enough to be granted privileged access to this material. In addition to the Group's correspondence, minutes of their meetings, membership records, and publicity material, there were witness statements from court cases, banners, and hundreds of photographs. Some of these photos were copies of images taken by sympathetic photojournalists, but many were photos taken by picketers outside the South African Embassy to record their protests or witness arrests.

We supplemented our analysis of City Group's archive with material from the AAM's archive at Oxford University, Norma Kitson's papers deposited at the Mayibuye Archive at the University of Western Cape,[6] Steven Kitson's personal papers (which were loaned to us by his sister, Amandla), and a number of news media archives.

It is no secret that, in addition to researching the Non-Stop Picket of the South African Embassy, we both participated in that protest – Gavin as a teenager and Helen, (initially) through her parents' involvement, as a younger child. Our research has consequently been informed by our auto-ethnographic reflections on our own experiences and participation there (Ellis 2004).

In many ways, then, it is not surprising that whenever we have presented aspects of this research at academic conferences, as much as colleagues want to hear our stories, they ask 'but how do you maintain critical distance?' Conversely, activists who we picketed alongside have worried that our analysis has become 'too academic' and 'too balanced'. These are people we shared intense experiences with; people we share common political assumptions with (even though we have all followed different trajectories in the intervening years); and people we trust. Balancing those close political and emotional bonds with our role as critical academic researchers can be fraught. We all want to share the good memories, but thinking through our failures and the things we could have done better or differently is a difficult and sensitive subject (for us all).

We endeavoured to retrieve and record the history of the Non-Stop Picket, collecting narratives from participants and considering what impact being 'non-stop against apartheid' has had on their (our) subsequent lives.

Recognising that the Non-Stop Picket only played an infinitesimal role in securing the release of Nelson Mandela and ending apartheid does not diminish the transformative effect that standing in solidarity outside the Embassy had on the lives of picketers. Neither does it deny that the existence and persistence of the Picket was a boost to those South Africans resisting apartheid who managed to hear about it. The Non-Stop Picket's solidarity had material effects in London, South Africa and elsewhere, and it helped generate new ways of understanding and engaging with the world for many of the people who came into contact with it. Understanding those processes and outcomes is what we are most interested in, rather than either measuring the Picket's 'success' or assessing the validity of its political stance.

This distance in time has allowed us to be a bit more detached, whilst also recognising that our critical detachment can never be anything other than partial. That might be a problem if we were trying to offer a hard-nosed, rational assessment of the veracity of the Picket's tactics, politics, and practices. However, because our research has been more interested in how and why this protest had such a dramatic and intense impact on the lives of many who participated in it, our shared experience of those events and their lingering legacies allows a deeper insight into those questions and a marker against which to interpret participants' memories and opinions.

An outline of what follows

The chapter that follows this introduction does three things: first, it narrates the story of the Kitson family and their role in forming the City Group; second, it examines how City Group developed the capacity to launch and sustain a Non-Stop Picket of the South African Embassy; and third, it articulates a theoretical framework for analysing their practice of anti-apartheid solidarity protest geographically.

Chapter 3 examines how the Non-Stop Picket inserted itself into the geography of Central London to quickly become a (seemingly) permanent feature of the city. It examines how its non-stop presence enabled such a diverse group of young people to become involved in its cause.

To spend time on the Non-Stop Picket was to experience time in a very particular way. For nearly four years, it *was* non-stop. It worked with (and sometimes against) the rhythms of urban life to practise its solidarity with the people of South Africa. Although the Picket was a constant presence (and was structured around a core set of activities), how it looked, how it functioned, and what it was like to be there changed throughout the day and across the week. Maintaining a 'non-stop' protest around an 'urgent' global issue required non-stop commitment from core activists, which was frequently hard to sustain. In addition to considering the temporalities of life on the Non-Stop Picket, Chapter 4 considers how that pace of activity fits with the experience of youth and the transition to adulthood. To maintain momentum, the Picket was structured around particular weekly rituals and an annual calendar of events. The Picket found ways of celebrating its longevity that served to recognise the commitment of existing activists and recruit new participants. In considering the way time passed and was marked on the picket, Chapter 4 examines the different rhythms of the protest – its daily, weekly, and annual cycles.

The Non-Stop Picket actively sought to disrupt the business of the South African Embassy. In response, the Embassy applied diplomatic pressure on the British government and the Metropolitan Police to curtail their protest. In this context, Chapter 5 examines the Picket's relationship with the police, examining some of the key points of contention between them and the various ways in which City Group defended their right to protest against apartheid in the location and manner of their choice. In particular, this chapter examines how, through a two-month campaign of civil disobedience, picketers regained the right to protest directly outside the Embassy gates after the Metropolitan Police once again forcibly moved them in May 1987. Through their non-violent, but confrontational, political stance, the young picketers learned to think and act against the (British) state, using their bodies in unruly ways.

City Group fostered a culture of direct action against the representatives of the apartheid regime (and their supporters) in Britain, which was expressed both on and off the Non-Stop Picket. Chapter 6 examines how picketers learned to be unruly in various ways through the direct actions they took in support of the economic and sporting boycotts of South Africa. In this context, we examine the practices through which City Group offered political and legal support to those arrested on its protests. These practices were particularly effective – of the more than 700 arrests associated with the Non-Stop Picket, over 90 per cent of cases were (eventually) won by the defendants.

Children and young people were central to sustaining the Non-Stop Picket. Through their shared commitment to anti-apartheid solidarity, young people from diverse backgrounds grew up together and learned to cope with the everyday pressures of youth. The anti-apartheid cause was not a backdrop to these young people's lives; they grew up *through* their political engagement. Chapter 7 argues that young activists' political commitments are always entangled with the everyday politics of youth; that (in the context of the Non-Stop Picket) to practise solidarity was also to develop competences and resources that contributed to the process of growing up. Although this chapter focusses on the experiences of teenagers and young adults, it also argues that 'youthfulness' and practices of 'growing up' are relational and not age specific. Several picketers who joined their protest in their thirties describe how their involvement with the social and political life of the Non-Stop Picket gave them opportunities to 'grow up' anew. There were also a small number of very committed elderly picketers, but few of them were still alive by the time we conducted our research.

The primary demand of the Non-Stop Picket was the unconditional release of Nelson Mandela. When Mandela was released in February 1990, the Non-Stop Picket had achieved its main goal and had to come to an end. Mandela's release was celebrated as a 'victory'; but for many participants, the abrupt end of the Picket felt like a loss. The protest that had become the focus of their lives for up to four years was gone, and the close bonds of comradeship they had developed there were threatened. Chapter 8 analyses activists' ambivalent experiences of victory and some of their reflections on how post-apartheid South Africa has turned out.

Finally, Chapter 9 examines the impact that participating in the Non-Stop Picket has had on the personal and political lives of former picketers (now that most have reached early middle-age). We explore how both the comradely relations of care that developed on the Picket and many of the constituent practices of non-stop picketing endure in their lives. Consistent with our earlier argument that young activists' political commitments are always entangled with the everyday politics of growing up, we suggest that youthful activism can be a valuable resource for socially engaged adulthood. This chapter makes a strong case for a social practices approach to activism that offers new possibilities for understanding the dynamic ways in which activist practices become bundled with other aspects of life and life-course transitions. In doing so, it extends the reach of recent debates about the transformative effects of practising solidarity. The book concludes by examining what lessons can be learned from the Non-Stop Picket for academics and activists interested in urban social movements, protest camps, young people's activism, and the history of the international movement against apartheid.

Notes

1 LBC/IRN, "Norma Kitson on South Africa", September 1986. http://bufvc.ac.uk/tvandradio/lbc/index.php/segment/0011900056007 (Accessed 02 Aug 2016).
2 Interview with Carol Brickley.
3 There is a parallel here with Craig Jeffrey's (2010) work on "timepass" and the politics of waiting for unemployed youth in India (although his study focussed on the experiences of more highly educated young men) or Southwood's (2011) thoughts on the "non-stop inertia" of precarious employment.
4 The year-long protest outside the South African Consulate in Washington, DC was not the only long-term emplaced 'embassy protest' that inspired City Group activists. Many were aware of the existence of the Aboriginal Tent Embassy for indigenous land rights in Canberra, Australia (Iveson 2016).
5 As Weeks (2007: 17) reviews, in 1983, 62 per cent of adults questioned by the British Social Attitudes survey disapproved of homosexual relationships. By 1987, at the height of the social and political panic around AIDS, the equivalent figure was 74 per cent. However, a 1988 Gallup Poll reported that 60 per cent of those questioned did not accept 'gay lifestyles'; however, amongst people aged under 25, 50 per cent of that age group were accepting of gay people (Weeks 2000: 171).
6 Thanks to Evan Smith for copying this material for us during his own, unrelated research in the Mayibuye Archive.

2 A non-stop protest in a non-stop world

From the inception of the Non-Stop Picket, the organisers pledged to remain on the pavement outside South Africa House in Trafalgar Square until Nelson Mandela was released from gaol. Starting on 19 April 1986, they did indeed stay there around the clock until (shortly after) Mandela walked free from prison in February 1990. Physically, the Non-Stop Picket consisted of little more than a handmade fabric banner and a few boxes of equipment; but, through its continuous presence in Central London and the intensity of its 'non-stop' protest, it came to feel anything but ephemeral. Unlike many contemporary protest camps (Brown et al. 2017), the non-stop picketers could not pitch tents and were seldom allowed to sit down outside the Embassy. One report in *City Limits* (London's erstwhile radical listings) magazine, published at the time of the Picket's 1,000th day, noted the seeming permanence of the Picket within the landscape of Central London:

> The non-stop picket has established itself as a part of London life since it was started by City AA on April 19 1986. It features on the tourist circuit. Guides on open-topped double-deckers point it out to their camera-wielding passengers as they cruise by. Travellers from all over the world take a photograph for the album, sign the 'Free Mandela' petition and throw a few coins in the collecting bucket.[1]

City Limits and its readership were sympathetic to (perhaps even fond of) the Non-Stop Picket. In December 1987, *City Limits'* readers had voted this continuous anti-apartheid protest as their 'Demo of the Year'. While the number of people who actually participated regularly in the Picket was relatively small, this award demonstrates the degree to which the Non-Stop Picket captured the imagination of a wider population. By December 1987, the Picket had stood on the pavement in Trafalgar Square, outside the South African Embassy, for 20 months. Thousands of people passed it every day, and the image of the Picket resonated with many Londoners. In some small way, the award from the readership of *City Limits* was recognition of this.

The City of London Anti-Apartheid Group (City Group), who organised the Non-Stop Picket, could not let this award pass without capitalizing on the opportunity for further media coverage. They arranged for the former South African political prisoner, David Kitson (the husband of Norma, who had co-founded the group), to be photographed on the Picket handing over the award to the protesters there. Winter on the Picket could be harsh, and each winter seemed to get harder. Good news stories, like this award, were always welcome at this time of year to boost morale and encourage City Group's members to continue physically supporting the Picket with their presence.

As well as thinking about how City Group practised solidarity with those fighting apartheid through the Non-Stop Picket, we are particularly interested in the eclectic community that was formed through the shared practice of picketing the South African Embassy. In some ways, it was the very thrown-togetherness (Massey 2005) of this 'motley crew'[2] that made the Non-Stop Picket such an interesting and vital place where the exchange of ideas between people from very different backgrounds and political perspectives fostered new forms of solidarity. Those who sustained the Picket for nearly four years were mostly (but not exclusively) young, but they came from a very diverse set of backgrounds, both in Britain and a large number of other countries around the world. For many people, intense friendships were formed through the Picket; but this was far from being a universal experience – some people always felt like outsiders on the Picket but kept going there anyway (frequently because of the intensity of their commitment to the anti-apartheid cause).

More than a quarter of a century after the Non-Stop Picket ended, there is no one who was a regular picketer who is not at least in their forties. They are, once again, dispersed across the globe. One of the questions this book addresses is: how did that intense experience of being on the Non-Stop Picket in the late 1980s impact and shape the subsequent lives of former picketers? This chapter outlines the story of the Kitson family and their part in the formation of City Group. It examines how, in the four years from its formation in 1982, City Group developed the capacity to launch and sustain a protest that would stand outside the South African Embassy continuously for nearly four years. Having sketched this history, we articulate a theoretical framework for geographically analysing how City Group practised their anti-apartheid solidarity through the Non-Stop Picket.

The Kitson family and the origins of City Group

David Kitson was not directly involved in the formation of City Group; but, without him, City Group would not have come into being (Brown and Yaffe 2013). David was one of the longest-serving white political prisoners in apartheid's jails until his release in 1984. His wife, Norma (herself an exiled member of the African National Congress (ANC)), and their children, Amandla and Steven, were central to the formation of City Group, which directly grew out of a campaign to free Steven after he was briefly detained (in January 1982) while visiting his father in jail in South Africa.[3] By retelling something of the family's story, we hope to explain how and why Norma Kitson and her children set out to build a campaign on the streets of London that would engage British youth in active anti-apartheid solidarity work.

David was a member of the South African Communist Party (SACP) and played a crucial role in the early days of the armed struggle against apartheid in the early 1960s, acting as a bomb instructor for Umkhonto we Sizwe (MK) when it was formed in December 1961. When almost the entire leadership of MK was arrested at Rivonia in July 1963, David was propelled into the four-strong replacement High Command of the organisation (Ellis 2013). He served a little more than four months in this role before he too was arrested. At his subsequent trial, he was jailed for 20 years for sabotage and membership of the Communist Party. He served almost the full term; eventually being released in May 1984.

David studied mechanical engineering in Durban in the 1940s. Upon graduation, he served as a sapper in the South African army, thereby learning to apply his training as an engineer in ways that would later benefit the anti-apartheid struggle. At the end of the Second World War, David moved to London, where his father had been born. There he worked as a draughtsman for de Havilland Aircraft. He played an active role in TASS (the Technical, Administrative, and Supervisory Section of the Amalgamated Engineering Union) and became Secretary of the Hornsey branch of the Communist Party. His union sponsored him to study at Ruskin College, Oxford for two years.

Whilst in London, David met and married Norma Cranko, a fellow South African who had been active in the Congress of Democrats, the white wing of the ANC-led Congress Alliance. Norma was in London preparing to participate in the 1955 Warsaw Youth Festival as part of the South African delegation (of ANC and SACP supporters). There she also joined the Communist Party. Following the birth of their son, Steven, they returned to South Africa in June 1959. Publicly, they returned to introduce Steven to his grandparents and raise him as a South African; but the length of their respective absences from the country meant that they were a valuable asset to the anti-apartheid struggle, and both Norma and David were instructed to join the emerging underground cell structures of Congress and the SACP (Kitson 1987).

Norma and David returned to South Africa, taking an active role in the struggle just as many other progressive whites were leaving the country. Although David was a Communist Party loyalist, he did not uncritically tow the 'party line' (Bundy 1989; Fine 1989; Ellis and Sechaba 1992).[4] In particular, David had developed an 'anti-imperialist' perspective on the role of South African revolutionaries and did not support the 'stagist' strategy of the SACP. He did not believe that a national democratic revolution should be prioritised over and ahead of a socialist one. Taken together, these two expressions of 'going against the flow' provoked tensions between the Kitsons and their comrades, and were to have consequences for them years later.

Following David's arrest, Norma was also detained for a period and endured persistent harassment from the police for as long as she remained in South Africa. In 1966, Norma decided to move Steven and Amandla to London for their safety. There, she was largely shunned by the exiled ANC and SACP community for many years but enjoyed support and friendship from many people she and David had known during their time in the Hornsey Branch of the Communist Party. Some of these people – Mary and Henry Barnett, Lena and Arthur Prior, and Rene Waller (names that will reappear throughout this book) – would later become stalwarts of City Group (and, later still, the Justice for Kitson Campaign[5]). In London, Norma divorced David, with his support, and married Sidney Cherfas, a fellow South African – although Norma and David remarried following his release from gaol.

Each December, from the age of 16, Steve used the holiday period to return to South Africa to visit David. In 1979, three white political prisoners escaped from Pretoria Local Prison (Keable 2012). In punishment, the authorities moved the other white political prisoners to death row. In the poor conditions of death row, David's health deteriorated rapidly, and the family became increasingly concerned about his well-being. On 6 January 1982, while visiting his father in gaol in Pretoria, Steve was detained by the South African authorities, accused of being an ANC courier and breaching prison security by sketching the institution.[6] Steve was tortured during his detention. Norma and her colleague at Red Lion Setters, Carol Brickley (a member of the Revolutionary Communist Group[7]), quickly mobilised everyone they could think of to demand Steve's freedom. The Free Steven Kitson Campaign was born, and he was released after six days. Within hours of phoning London with news of his release, Steve's aunt, Joan Weinberg (Norma's older sister), was murdered in her flat in Johannesburg.[8] With Norma and the children in London, Joan had been David's most frequent visitor throughout his imprisonment. Her killers were never found; indeed, they were never sought.

During its brief existence, the Free Steven Kitson Campaign drew scores of new people into anti-apartheid campaigning for the first time. In order not to lose this momentum, it was decided to transform the campaign into City of London Anti-Apartheid Group (as a local branch of the national Anti-Apartheid Movement (AAM)[9]).

David's health continued to deteriorate, and his family became increasingly frustrated by the refusal of the British Foreign Office to intervene on his behalf and secure a move to less brutal conditions. In August 1982, on David's 63rd birthday, City Group launched a non-stop picket (24 hours a day, seven days a week) of the South African Embassy in London, calling for his release (Stevens 2014). On 8 November 1982, David Kitson and the other prisoners were moved from death row to more suitable prison accommodation (Kitson 1987). The picket organised a victory rally to celebrate. Having won its primary demand, and in the face of hostility from the London ANC – who had stated they were only prepared to sanction a family campaign for David's release but then caught the family in a catch-22 situation by saying that they and the AAM could only support campaigns for *all* political prisoners, (Kitson 1987: 269), Norma Kitson and other leading members of City Group proposed that the picket should end. The suggestion was not universally popular – many of the new, younger picketers wanted to keep going. City Group organised democratically and empowered its membership to decide its campaigning strategy. On 18 November, by a slim majority, City Group's membership voted to end the first non-stop picket. It had lasted for 86 days and nights. With this apparent victory under their belts, City Group continued to hold weekly protests outside the Embassy.

City Group's confrontational approach to anti-apartheid campaigning won few friends amongst the exiled ANC members in London and their supporters in the British Anti-Apartheid Movement (for reasons that will be explained later in the chapter). When David arrived in London following his release in May 1984, the SACP instructed him to denounce Norma and call for City Group to be disbanded. Unsurprisingly, he refused. His principled refusal to disown the group that his wife and children had helped to found cost him dearly. By November 1984, both Norma and David had been expelled from the SACP and suspended from the ANC[10] (although, as David was fond of pointing out, with slight bemusement, he had technically never been a member of the ANC[11]). In the process, he also found that the funding for the post at Ruskin College, the trade union college in Oxford, which had been promised to him during his imprisonment, had been withdrawn by Ken Gill (the General Secretary of his union, TASS (now part of Unite)). David lost his main source of income as a result. In his official history of the British Anti-Apartheid Movement, Fieldhouse (2005: 409) acknowledges that the union's decision to rescind the funding for David Kitson's post at Ruskin was "not a very edifying episode."

Norma and David always maintained that there were deeper politics to their treatment: that David's release, after 20 years, was a threat and embarrassment to those white SACP and ANC members who had fled South Africa (against the instructions of the SACP Central Committee) while he was still working in the underground. That he had long-standing political differences with Joe Slovo[12] and others over revolutionary strategy was further fuel for this conflict. These political differences were real, but they were also manipulated by South African Military Intelligence at the time – the two ANC leaders who suspended the Kitsons, Solly Smith and Francis Meli, were both later revealed to have been South African agents (Trewhela 1995; Bell with Ntsebeza 2003). Through City Group and the Justice for Kitson Campaign, David fought hard to clear his name, but with few victories. Following their release from gaol, Nelson Mandela and Walter Sisulu called for the suspension of the Kitsons from the ANC to be lifted in the early 1990s, but David and Norma were never incorporated back into ANC structures and politics.

Developing practices and capacity

The longevity of the 1986–1990 Non-Stop Picket would probably not have been achievable were it not for the lessons learned in 1982. The non-stop picket for David Kitson had lasted 86 days, but the culture of solidarity and organising that it had fostered served City Group well as they continued campaigning against apartheid for another 12 years. Here, we chart how and why City Group developed the capacity to launch the Non-Stop Picket in April 1986 that was determined (and able) and remained until Mandela was released from gaol. The Non-Stop Picket lasted for nearly four years, but that duration could not have been anticipated at its beginning.

The first continuous picket had many similarities with the Non-Stop Picket that started four years later. Much of what became City Group's approach to solidarity work, including the use of strong, colourful visual material, and a lively, noisy culture of song and chanting, developed during the picket in 1982. There were crucial differences, though. In 1982, the night shift slept on pavement while people took turns on security detail. Thanks to the intervention of Lord Gifford, the police allowed a small cluster of chairs – 'the lounge' – to be used on the picket. Over the course of the 86-day picket, City Group grew and drew in new groups of supporters. As Norma Kitson described (in relation to the picket for David Kitson in 1982),

> At first we were a thin line of City Group members and friends of our family, but soon people passing by, students from colleges, schools and dole queues joined us, and stayed. Many came as shy individuals, knowing little about South Africa. Amandla's powerful voice rang out, chanting slogans and calling on people to sign the petition. Steven and I taught liberation songs, and we soon became a family.
> (Kitson 1987: 270)

In her autobiography, Norma acknowledged that when the 1982 picket started, she was scared of many of the young people it attracted, and she favoured older picketers (with a proven track record of campaigning) to take the lead in organising the protest. But, she also celebrated the way in which the young picketers (often teenagers) quickly educated themselves about apartheid, took political initiative, and took responsibility for stewarding the picket.

Ann Elliot was one of those older, proven campaigners who was involved in City Group from its beginning. She celebrated the 'exhilaration' she felt on the picket (compared to her previous experience of the British Left), explaining that this was:

> because of the particular qualities of this demonstration which, I will say, were bought to it by Norma Kitson and her experience of struggle in South Africa. I think I personally, and my political tendency, owe

a great deal to Norma's interpretation of the South African struggle in the British situation. This involved things like having children and women very much to the fore, having colourful and noisy demonstrations and having singing. This was a completely different tradition from the one I sort of breached from my earliest days, which was trade union meetings which undoubtedly was roomfuls of grey men in grey suits full of smoke.[13]

Elsewhere in her interview, Ann links Norma Kitson's political analysis to the ways in which she encouraged City Group to perform its solidarity. The form and content of anti-apartheid protest was linked for Norma. Her understanding of the situation in South Africa led her to appreciate the importance of noisy demonstrations to disrupt the work of apartheid's representatives in London. Because she was committed to mobilising as many people as possible to anti-apartheid work, she recognised that demonstrations should be visible, lively and attractive – that they should be colourful and noisy with singing. Norma's vivacious approach to solidarity is contrasted to the grey men of the British Left in the preceding (and concurrent) period. In this respect, clearly, Norma's approach was not unique (even if it was still relatively unusual in British anti-apartheid work) – in the early and mid-1980s, women peace activists at Greenham Common playfully reworked many assumptions about who could take political action, as well as what that might look and feel like (Cresswell 1994; Roseneil 2000; Feigenbaum 2013).

Ann passionately articulates the ways in which the appearance and performance of City Group's pickets of the Embassy were connected to their modes of conduct. For her, none of these qualities were simply superficial; they stemmed from (and encouraged engagement with) a particular way of analysing the struggle against apartheid. For her, Norma Kitson was one of the key figures in developing this approach as a "new step forward in the politics of protest in Britain." It is not surprising that while she inspired many people, Norma's passion and creativity also upset the bearers of long-established traditions on the British Left (and within the progressive South Africa diaspora in London). From the 86-day picket in 1982 onwards, the relationship between Norma Kitson and the leadership of the London ANC was tense and then deteriorated. Given the centrality of ANC to the leadership of the Anti-Apartheid Movement in Britain, the relationship between City Group and the national movement also came under strain.

In June 1984, South African President PW Botha was expected in Britain for talks with Margaret Thatcher. His tour of Europe that summer was intended to promote 'constructive engagement' with the apartheid regime (rather than sanctions) and stave off a major political crisis for South Africa (Byrd 1988; Guelke 2005). The national Anti-Apartheid Movement planned various protests in response. City Group called a week-long continuous picket of the South African Embassy from 26 May until 1 June, immediately prior to Botha's arrival.

Just before Botha's visit to Britain, on 25 May 1984, a volunteer from the national Anti-Apartheid Movement's office threw paint over the main entrance to the South African Embassy. On 1 June, four members of the National Union of Mineworkers deposited a pile of coal on the Embassy's doorstep in protest at the importation of South African coal in an attempt to break the miners' strike (Rutledge and Wright 1985). Given the heightened security concerns surrounding Botha's visit, the Metropolitan Police took the opportunity to ban all protests outside the South African Embassy, with effect from Friday 8 June 1984. A press statement, issued by City Group on 10 June, explained these events:

> On 5 June Superintendent Dark of Cannon Row informed our Convenor that the City AA picket scheduled for Friday 8 June, 5.30–7.30 p.m. was banned. He claimed that all pickets and demonstrations in the vicinity were being banned due to a banquet for Reagan, Thatcher & Co being held the same evening at the National Portrait Gallery. By the evening of 5 June, following protests by MPs, councillors and others, permission was given for a picket in Duncannon Street-near the National Portrait Gallery-but not outside South Africa House. On Thursday 7 June, despite consistent misinformation, spread by Scotland Yard to the Press, it became clear that what was at issue was a permanent ban on City AA's pickets and not concern for the Reagan/Thatcher banquet. The City Group convenor and Richard Balfe MEP met Commander Howlett on 7 June. [...]
>
> Commander Howlett informed us that this was his decision based on his own personal interpretation of the Vienna Convention catering for the peace and dignity of embassies. In his opinion any picket or demonstration outside any embassy would be in breach of the Vienna Convention. He admitted to having consulted no-one in arriving at this decision.[14]

The press release continued by explaining City Group's response to the ban:

> City AA was prepared to resist this attack. On 8 June over 200 people representing a wide range of organisations assembled at Duncannon Street. Among the demonstrators were Ernie Roberts MP, Jeremy Corbyn MP, Rodney Bickerstaffe of NUPE, and Peter Hain.
>
> At approximately 6 p.m. a group of 18 demonstrators crossed the road and assembled in front of the Embassy. They began to sing, chant slogans, distribute leaflets and collect signatures. Within 5 or 6 minutes a massive police cordon surrounded the peaceful picketers. They were all arrested without being told why and bundled into waiting police vans.[15]

Half an hour later, another six protesters crossed the road and were also arrested. The 24 anti-apartheid demonstrators were taken to Albany Street Police Station nearby. They were all released without charge after only a couple of hours.

In response to this ban on demonstrations, City Group convened the South African Embassy Picket Campaign (SAEPC) to win back the right to protest against apartheid directly in front of South Africa House (Stevens 2014). The defiance of the ban continued, with people risking arrest each Friday. On 15 June, a senior police officer, Commander Howlett met with Mike Terry from the national Anti-Apartheid Movement to discuss the ban. In a witness statement[16] (dated 6 July), Commander Howlett described this meeting in the following terms:

> On Friday, 15th June 1984, at 10.15 a.m., together with Chief Superintendent Richards and Inspector Menear, I met, at my request, in my office, Mr Mike Terry and Miss Cate Clark of the Anti-Apartheid Movement. I explained the change in police policy and the reasons for it to them. A note of the meeting has been kept by the police. Mr Terry expressed their opposition to the change and indicated that steps to alter that policy would be taken and that his organisation intended that those steps be legal as his movement did not seek confrontation.

When a representative of the SAEPC met with the police later that day, they were advised to follow the lead of the AAM in avoiding confrontation. The SAEPC did not take that advice – that evening, 26 protesters crossed Duncannon Street to break the ban. This time, they were charged with obstructing the police in the cause of their duties (in this context, their duties were said to be protecting the 'peace and dignity' of the South African Embassy in line with the Vienna Convention (1961)).

On Friday 22 June, David Kitson (who had, by then, been released from jail in South Africa) watched 22 anti-apartheid activists, including his son Steve, be arrested for defying the ban. As the defendants began appearing at Bow Street Magistrates Court the following week, strict bail conditions were imposed, preventing the defendants from standing on Morley's Hill, the pavement in front of South Africa House. By 16 July, at least 137 people had been arrested as a result of the SAEPC and five had been imprisoned for a week for breach of the bail conditions forbidding them from demonstrating within 30 yards of the Embassy.[17] During the course of a 24-hour protest on 21/22 July 1984, five councillors and three Members of Parliament (MPs) – Tony Banks, Stuart Holland, and Jeremy Corbyn – were arrested for breaking the ban.

In an act of solidarity with its arrested councillors and the SAEPC, Camden Council took a legal opinion, on the legality of the ban, from Stephen Sedley, Queen's Counsel (QC).[18] His view was:

> The Vienna Convention 1961, insofar as it is incorporated in the Diplomatic Privileges Act 1964, adds nothing to the duties, nor therefore to the powers, of the police in this regard. It follows, if I am right, that their action in declining to allow continuance of the demonstrations on

the pavement outside South Africa House was an abuse of their powers, and that they were not executing any duty known to the law when they cleared the pavement and arrested persons who remained there.

One of the City Group defendants, Richard Roques (a former school friend of Steve Kitson who became a supporter of the Revolutionary Communist Group (RCG) and City Group's Treasurer), agreed to stand as a test case. On 1 August 1984, at Bow Street Magistrates Court, the Chief Stipendiary Magistrate for Central London, Mr Hopkins, dismissed the charges against Richard. A week later, in light of this decision, the police dropped the charges against the other 136 protesters who had been arrested during the campaign (including the five who had been imprisoned). The victory for the SAEPC was an embarrassment for the national Anti-Apartheid Movement. The Executive Committee of the AAM had issued a statement on 10 July 1984.[19] Their statement outlined a series of meetings they had arranged with senior police officers to try to negotiate an end to the ban:

> It is the view of the Executive that until the above process has been exhausted and the EC has met to further assess the situation, the AAM and its local groups should not demonstrate immediately in front of South Africa House.
>
> The AAM has been approached by a number of local groups and others seeking advice on the '"South African Embassy Picket Campaign 1984". The AAM EC does not support their approach, believes that it damages the prospect of achieving a removal of the ban, and therefore asks its members and supporters not to participate in this campaign.

The AAM was wrong to be so cautious. Although City Group and the SAEPC won back the right to protest outside the South African Embassy through their civil disobedience, the AAM would make them pay for their victory. City Group's defiance of the AAM Executive Committee's guidance accelerated the deterioration of the (already strained) relationship between the national movement and its 'hooligan element' in the City of London.[20] The conduct of the SAEPC remained a contentious issue with the AAM and directly contributed to City Group's 'disaffiliation' from the national movement the following year. City Group remained committed to taking direct action against apartheid and civil disobedience to defend the right to protest; not just in South Africa, but also in Britain (see Chapter 5).

On 23 February 1985, City Group ceased to be an officially recognised local group of the national Anti-Apartheid Movement. This was the culmination of more than two years of growing political tensions between the leadership of the national movement and City Group about how to build an effective solidarity movement.

While the National Committee of the Anti-Apartheid Movement met in Cardiff to decide City Group's fate, City Group members were holding a

24-hour picket outside the South African Embassy in London. The irony of this comparison was drawn out in an article published in *Fight Racism, Fight Imperialism* (the RCG's monthly newspaper) at the time:

> There could be no greater contrast between the singing youth, black and white, employed and unemployed, on the 24 hour picket outside South Africa House organised by the City of London Anti-Apartheid Group on 22/23 February, and the tired, cynical and manipulative body of AAM committee men and women who met in Cardiff on the same day to rid the movement of City Group's influence and enthusiasm.[21]

The AAM leadership repeatedly asked City Group to give an account of its actions and comply with the AAM's constitution (Brickley et al. 1986; Fieldhouse 2005). Ultimately, the reason given for City Group's removal as a local AAM group in February 1985 was a weak, technical one – that it failed to restrict its activities to the boundaries of the City of London and did not restrict its membership to those living and working within the City of London. It is certainly true that most of City Group's activities took place in other areas of Central London, and it recruited its members far and wide. However, technically none of these actions represented a breach of the AAM constitution. As Fieldhouse (2005: 220) acknowledges:

> Although some of these complaints were justified and were aggravating, none of them offended against any written rules of the Movement, nor would they have caused much concern had they not been coupled with an unacceptable ideology.

He continues:

> It is also true that at times the officers and executive [of AAM] were not above inventing rules to impose on CLAAG, for example that they should confine their activities to a particular geographical area or refrain from distributing non-AAM literature at demonstrations. There were no such written rules and many other groups overstepped them.
> (Fieldhouse 2005: 226)

City Group made exactly this point loudly and repeatedly in contesting the National Committee's decision at the time. The subtleties of language are important in such disputes: City Group maintained that they were 'expelled' from the national AAM; while, to this day, some commentators close to the AAM leadership continue to insist that they were merely 'disaffiliated' as a local group.[22] At the heart of this semantic dispute is the question of the AAM's constitution – they did not have grounds to expel City Group from the national movement; all they could do was cease to recognise them as a local group within the movement.[23]

Exactly six months to the day before the Non-Stop Picket started, City Group led a spectacular protest outside the South African Embassy in Trafalgar Square. On 19 October 1985,[24] 322 people were arrested for blocking the road directly in front of the Embassy.

In the weeks preceding the demonstration, City Group had circulated a call to surround the South African Embassy. On the day, City Group's numbers were swollen by students participating in a National Union of Students (NUS) demonstration against apartheid in Trafalgar Square. Over the course of the afternoon, hundreds of students relocated from the centre of Trafalgar Square to protest directly outside the South African Embassy and eventually blocked the road in front of it.

Raw footage of the protest held on the Independent Television News (ITN) Archive[25] shows hundreds of young people blocking the road. Both the pavement outside the Embassy and the roadway itself were packed full of people. At the start of the clip, some are standing in the road, and some are sitting. Two buses are caught in the crowd, prevented from moving. A familiar City Group chant is heard coming from the crowd – "Close down the nest of spies! Stop the murder! Stop the lies!"[26] As the police move in to make arrests and clear the road, student protesters make their bodies limp and are carried away. More seasoned City Group activists continue chanting for the release of Nelson Mandela as they are arrested, and a legal observer busily weaves between the police trying to record the names of arrestees.

If, in hindsight, many protesters remembered this as a peak event, in the aftermath of the protest, leading City Group activists offered a more balanced assessment. The following is taken from a report written by Andy Higginbottom, City Group's Secretary, for their Annual General Meeting the following February (at which, incidentally, the idea for a new Non-Stop Picket was first proposed):

> It should be noted that the success of 19 October, which was registered in the jubilation of almost all who took part (bar the police) was not just because we got it right on the day. In fact, at a certain level of assessment it could be quite self-critical of City AA's performance on the day. Far too many of our people were arrested early on, leading to a shortage of stewards and organisers, for instance. But this would be missing the main point. We got it right on the day because the political conditions in Britain correspond to City AA's overall approach and methods. 19 October was a vindication of our pushing forward and involving thousands in street protests, it was a sharp rebuttal to those who put themselves forward as leaders in order to stop the escalation of protests at all costs. Those who are hostile to City AA immediately began to minimise its significance or leave out the fact that City Group organised the event.
>
> The 19 October events showed that direct action protest to close the Embassy is possible, is legitimate and can be extremely embarrassing to the British government and their racist allies, as well as to South Africa itself.[27]

This assessment has some validity – the sit down protest on 19 October 1985 demonstrated that hundreds of young people were prepared to take direct action and risk arrest in pursuit of the closure of the South African Embassy. Although City Group probably did not mobilise the majority of participants in the 19 October protest, it is clear that City Group's vision of what anti-apartheid protest could be (and their practical intervention amongst the demonstrators) was decisive on the day. The events that afternoon helped consolidate City Group's reputation for direct action against apartheid in a way that, and six months later, would make the launch of the Non-Stop Picket viable.

Geographies of solidarity

The end of apartheid was brought about by a combination of 'internal' factors within South Africa, such as the township uprisings and trade union militancy of the mid-1980s that made the country 'ungovernable' to some extent, creating splits within the South African ruling class; and, 'external' factors such as South Africa's military defeat in Angola and the peace accords that followed from it, the end of the Cold War, and the pressure of international sanctions and other outcomes of the international solidarity movement (Jaster 1990; Bond 2000; Guelke 2005; Westad 2005).

The international campaign against apartheid in South Africa, running from the late 1950s until the end of white minority rule in 1994, was one of the most widespread and sustained social movements of the last century. Thörn (2006, 2009) has argued that the international anti-apartheid campaign was one of the first really global social movements. The international solidarity movement consisted of networks of organisations and flows of individuals, ideas, policies, and activist tactics that were highly mobile, crossing geopolitical, ideological, and cultural borders (Thörn 2009: 418–9). These (inter)national movements drew together churches, trade unions, student groups, political parties (primarily on the Left and Centre-Left), social movements, and anti-colonial networks; each constituency carried with it its own historical legacies, ideological commitments, and organisational traditions (De Gruchy 1986; Gurney 2000; Thörn 2009: 434). Thörn (2009: 419) argues that, in weaving together an alliance between these disparate traditions, the Anti-Apartheid Movement "contributed to the construction of a transnational political culture that was a part of wider, complex and multi-layered processes of political globalisation in the post-war era," including postcolonial struggles, the Cold War, and, in later years, emerging experiences of neoliberal globalisation.

Our analysis of the Non-Stop Picket positions that protest within the longer history of anti-apartheid solidarity work in Britain and internationally. We examine how the Non-Stop Picket was located socially, politically, and geographically at multiple spatial scales simultaneously. It existed as a protest on a particular patch of pavement, outside a specific building, and in a famous square in the centre of London. It existed as a node within an international network of anti-apartheid, anti-racist, and anti-imperialist social movements. Through its solidarity, the Non-Stop Picket sought to disrupt existing political and economic flows between Britain and South Africa and to help bring new geopolitical arrangements into being. Both the Non-Stop Picket (as a protest) and the solidarity it practised were inherently geographical. In attending to the geographies of the Picket, we contribute to geographical debates about the spatialities of solidarity (Koopman 2008; Featherstone 2012; Routledge 2012). We argue that too often, in the broader social sciences, discussions of political solidarity overlook the range of practices through which solidarity is mobilised and enacted. International solidarity is frequently presented as an asymmetrical flow of assistance, travelling from one place to another. In contrast, we argue that relations of solidarity can travel in more than one direction, building complex webs of mutuality and reciprocity over time (Brown and Yaffe 2014; Kelliher 2017). While the political framing used to mobilise international solidarity is important, we argue that this does more than just articulate connections between distant places; it also shapes the practices through which solidarity is performed and the form that solidarity takes. We argue that it is vital to pay greater attention to the practices through which these solidarities are enacted in key sites.

Defining solidarity as "a relation forged through political struggle which seeks to challenge forms of oppression" Featherstone (2012: 5) has argued that the concept "has rarely been the subject of sustained theorization, reflection and investigation." Nevertheless, political philosophers have renewed their interest in exploring relations of solidarity in recent years (Gould 2007; Scholz 2008). This body of writing largely attends to the ethical commitments that might inspire acts of solidarity rather than the practices through which solidarity is generated, mobilised, and practised. One useful insight offered by Scholz (2008: 34) is that, unlike the forms of social and civic solidarity identified by Durkheim and other sociologists, political solidarity tends to be performed by collectives that share a perception of an injustice but are not necessarily unified by "shared attributes, location, or even shared interests." Scholz (2008: 56) notes:

> To be in solidarity with those who suffer is to work for social change to alter the conditions that create that suffering, but simultaneously those in solidarity may need to respond directly to the concrete needs of others and help to alleviate suffering. That is, political solidarity heeds a call for aid in multiple ways – the efforts of those people with the tools and means to assist others in distress might also be fulfilling the moral relation of political solidarity.

Whilst solidarity does frequently respond to a 'call for aid', there are two limitations to the way in which Scholz conceptualises acts of political solidarity. First, although correct in identifying the necessity of concrete action to enact social change to alleviate inequalities or oppression, her interest in the moral obligation to act overlooks any concrete examination of the range of actions that can be undertaken as acts of solidarity in specific circumstances. Second, her choice of language – the 'alleviation of suffering' – comes close to reducing political solidarity to acts of humanitarian assistance and denies the possibility of more entangled relations of reciprocity and mutual solidarity that might seek to enact concrete social change in more than one location.

Unsurprisingly, geographers have paid specific attention to the spatial relations of solidarity and the ways in which power and privilege are entangled in these relations. Writing specifically about the challenges of sustaining progressive forms of solidarity within the alter-globalisation movement of the early 2000s, Massey (2008: 313) observed that "[o]ne of the problems that all such campaigns face is how to establish solidarity between different places and different struggles." She suggests that across physical geographical distance and unequal access to resources, campaigns in different localities commonly seek to articulate and mobilise solidarity by stressing the ways in which those places are connected. The connections between distant places can be used to generate political solidarity in one of (at least) three ways. First, the political and economic relations connecting those places are identified as the root of the problem. Second, similarities between the places are identified. Or, third, a common enemy that affects both places is identified as their connection. Although she appears to suggest these as alternatives, we believe multiple different configurations of these analyses are often mobilised to frame (the need for) solidarity.

City Group framed its solidarity in terms of the connections, dating back to British colonial rule in Southern Africa, between the British and South African ruling classes, arguing that the same forces that benefited from apartheid also benefited from racism in Britain (Brickley 1985; Brickley et al. 1986). In fact, the group deployed aspects of each of Massey's three approaches to fostering solidarity – it identified British political and economic support for South Africa as a key issue (the material flows between the two nations); it identified common experiences of racism and oppression in both countries; and it identified a common enemy in the form of the capitalist class. Through political framings such as this, it is possible to identify the centrality of geographical relations to the politics of solidarity.

What is crucial here is the act of 'rethinking' the political possibilities for how and why places are connected. Whilst humanitarian solidarity may simply seek to sooth the worst expressions of global inequalities, campaigns for political solidarity seek to intervene in the connections and flows between places, refusing to participate in the reproduction of inequalities and oppression, disrupting and attempting to change "some of the dominant, more settled trajectories" (Massey 2008: 323). In making this point, Massey provides an important reminder that power and responsibility for inequality and injustice can be "distributed along long chains of command" (2008: 323) and, consequently, the sites at which acts of solidarity can be practised to disrupt these flows are multiple. In this respect, she argues, local and particular struggles are still crucial to transnational solidarity networks.

If Massey (2008; see also, Featherstone and Painter 2013) was mostly concerned with how different kinds of spatial relationships are understood and presented in the articulation of transnational solidarity, Featherstone (2012) considers both how solidarity is performed through spatial practices and what its effects might be. He explores solidarity as a 'transformative relation', arguing that the practice of extending solidarity across distance and difference plays an important role in "the active creation of new ways of relating" (Featherstone 2012: 5) and "new ways of configuring political relations and spaces" (2012: 6). While Massey (2008) stresses the importance of communication and meeting places in the articulation and generation of transnational solidarity, Featherstone (2012: 6) extends this argument by acknowledging that solidarities are "shaped through diverse exchanges, contacts and linkages."

Of particular relevance to the case of international anti-apartheid solidarity, Featherstone notes that one key practice of solidarity is to apply 'pressure from without' on the perpetrators of inequality and oppression. He argues that "[t]his 'pressure from without' can reshape the terrain of what is politically possible and what counts or is recognised as political. This contestation produces new ways of generating political community and different ways of shaping relations between places" (Featherstone 2012: 7). For Featherstone (2012: 7), then, solidarities are about more than the activation of relations between pre-existing communities and collectives. Collectives come into being through the act of mobilising solidarity, and a key practice for many solidarity campaigns is the act of engaging not only with those already committed to the cause, but the extension of the solidarity collective through the engagement of new people. Given its location on a busy thoroughfare in the heart of Central London, the Non-Stop Picket actively sought to engage passing members of the public and devised particular practices to invite their participation in support of their cause and in support of the picketers (see Chapter 3).

Central to Featherstone's (2012: 16) theorization of solidarities is the suggestion that, by "[l]ocating solidarities as world-making processes, by tracing the geographies they shape, contest, [and] rework," a greater understanding can be gained about "their productiveness and agency." We share this ambition and believe geographers can make a significant contribution to the study of transnational solidarities by attending to both the spatial relations and the embodied spatial practices through which they are enacted (Brown and Yaffe 2014). Our work extends this approach in at least three significant ways. First, we argue that the discursive ways in which solidarity is framed politically cannot be separated from how it is performed. Second, we suggest that more attention needs to be paid not only to how ideas travel through key nodes in the articulation of transnational solidarities, but also to the micro-politics of the practices through which these solidarities are enacted (Davies 2012). Finally, we develop Kelliher's (2017) attention to the ways in which solidarity can be mutual and intersectional, enabling activists to support each other (locally and at a distance); not just because of their commonalities, but also with due acknowledgement of their differences. In the case of the Non-Stop Picket, this meant both drawing the links between the operation of apartheid in South Africa and racism in Britain, but also generating solidarity to support those activists who were criminalised and harassed in the course of their solidarity activism in London.

As Featherstone (2012: 30) has astutely recognised, "the geographies of power through which solidarities are fashioned can bear in significant ways on the political alternatives they generate." For City Group, this not only meant intervening in the uneven power geometries through which apartheid was sustained, but also operating within a broader solidarity movement within which the ANC and its allies were the dominant political force in determining who should receive solidarity. This had a significant impact on how City Group framed and practised its solidarity, as shown in the following sections.

Practising solidarity

In the four years between Steve Kitson's detention in South Africa and the launch of the Non-Stop Picket, City Group had grown from a loose group of friends and colleagues of the Kitson family to one of the largest local groups within the Anti-Apartheid Movement in Britain (prior to its expulsion, at least). Through the 86-day non-stop picket for David Kitson, and four years of weekly pickets of the South African Embassy, the group had developed a repertoire of contentious actions, a suite of organisational practices, and a distinctive way of framing the need for a militant anti-apartheid solidarity movement on the streets of London. These practices helped recruit a large number of (primarily) young people to anti-apartheid campaigning, convince many of them of the validity of taking direct action and risking arrest to advance their campaigning, and develop a core group of them to play leading roles in running the Non-Stop Picket once it was launched (Brown and Yaffe 2016).

In this book, we outline a theoretical framework for understanding the geographies of solidarity and rethinking young people's political agency, activism, and engagement in geopolitics. Geographers (Massey 2008; Featherstone 2012) have theorized political solidarity as a spatial relationship that reworks the connections between places and attempts to forge new relationships between them. The Non-Stop Picket did just that – City Group articulated a politics of solidarity that sought to draw connections between the struggle against apartheid in South Africa and political struggles in Britain.

We extend this argument by theorizing solidarity as a social practice (or rather a bundle of related practices). On the Non-Stop Picket, these practices included, but were not limited to: educating the public about apartheid, mobilising support for action against apartheid (itself a bundle of various campaigning practices), collecting material aid for those resisting apartheid, and fostering anti-racist cultures of resistance. Drawing on Shove et al.'s (2012) conceptualisation of social practice theory, we examine these social practices as developing out of a dynamic relationship between materials (e.g. objects and technologies), competences (e.g. skills and techniques), and meanings (e.g. symbolic understandings, motivations). This moves our analysis beyond the traditional tropes of social movement theory (Jasper 1997; Della Porta and Diani 1999; Tilly and Tarrow 2007; Buechler 2011). Our approach is particularly attuned to considering the geography of when and where protests take place – for example, the Non-Stop Picket's location in Trafalgar Square not only had symbolic *meaning* for its solidarity, it also became a *material* resource that enabled the participation of a wide range of people.

We develop our use of social practice theory in one further, innovative way to think specifically about young people's activism. We argue that *both* solidarity *and* growing up are (bundled) social practices and consider how, in the context of the Non-Stop Picket, these two social practices came to be entangled, sharing common elements. Featherstone (2012) has argued that acting in solidarity has transformative effects on solidarity activists (as well as those who receive their support), because it can change how people understand their place in the world. We extend this approach further, by examining how solidarity, as a social practice, may also contribute to young activists' transition to adulthood (see Chapter 7).

City Group's solidarity operated at several scales simultaneously and was orientated in multiple directions. City Group understood its role in opposing apartheid as subordinate to the actions of the black majority in South Africa and their popular organisations. They framed their role as a solidarity organisation in Britain as to expose, target, and disrupt British economic, political, and diplomatic links with apartheid South Africa. They believed this could not be effective without also challenging systemic racism in Britain (Brickley et al. 1986; Williams 2012). Alongside these tasks, their role was to offer political, moral, practical, and material support to all those resisting apartheid in South Africa. Collectively, the group embodied these aims through their conduct of the Non-Stop Picket and associated campaigns. The Non-Stop Picket was formed with three central demands: (1) the unconditional release of Nelson Mandela, (2) the release of *all* political prisoners in apartheid gaols, and (3) the closure of the South African Embassy in London. These demands were carefully formulated. On the face of it, these demands (or, at least the first two) seemed relatively uncontroversial, but they created space for the articulation of a politics that differed from that of the national Anti-Apartheid Movement in Britain in crucial ways. First, the apparently innocuous call for the release of *all* political prisoners was significant – without directly saying so, this was a coded reference to what City Group referred to as it's 'non-sectarian' approach to solidarity work – namely that the group supported all progressive anti-apartheid tendencies in South Africa, not just the ANC. Officially this was also the position of the AAM. While, at times, they called for the release of 'all political prisoners', in practice, the AAM seldom actively campaigned for non-ANC cadre (Fieldhouse 2005: 115–120; Bell 2009; Klein 2009: 458).

We are cautious of presenting too strong a distinction between the political framing of City Group's solidarity and the practical enactment of those principles and demands. The group's politics developed through praxis. Ideas, demands, and policy positions were formulated through the group's practice: they were debated formally at meetings and informally on the Picket, they were presented in literature that was circulated on protests, and they changed over time.

The demand for the closure of the South African Embassy served two purposes: first, a focus on what the group termed 'British collaboration with apartheid' and second, an opening to discuss the links between apartheid in South Africa and racism in Britain (Williams 2015). As Carol Brickley, City Group's Convenor, said in her speech at the launch of the Non-Stop Picket:

> Britain is up to its neck in apartheid and that's why we're here today, that's why we're making our protest and that's why we're going to stay here.[28]

The South African Embassy was chosen as the prime focus for City Group's political campaigning due to its central, highly visible location, and for its symbolic place as the main representation of apartheid in Britain. In this way, City Group went about "making links [between distant places] localizable and contestable" (Featherstone 2012: 18). They sought to do more than just bear witness to the crimes of apartheid by standing outside the Embassy; they sought to protest there in such a way that they disrupted its normal functioning. The Non-Stop Picket was a form of emplaced, durational, direct action. The leadership of City Group was well aware that this disruptive and confrontational expression of protest would provoke a reaction from the Metropolitan Police. They used arrests, episodes of police harassment, and resulting court cases politically as evidence of 'British collaboration with apartheid' and as a vector for highlighting the links between racism in Britain and apartheid in South Africa.

If their aim was to dismantle the colonial relationship between Britain and South Africa, then City Group understood that the struggle against apartheid (internationally) had to be led by South Africans themselves. They believed that only the resistance of the South African people could end apartheid and that their primary objective, as solidarity activists, was to support that *and* to challenge those forces in Britain that perpetuated and profited from apartheid in South Africa. This formulation of the subordinate role of an international solidarity group developed as (and through) the Group's critique of the work of the national Anti-Apartheid Movement and its acceptance of the ANC as the 'sole, legitimate representatives' (Thomas 1996: 234) of the black majority in South Africa (but it was undoubtedly also influenced by the RCG's development of a similar ethics of solidarity in the 1970s in relation to the Irish Republican movement (Reed 1984)). Whilst the AAM saw itself as subordinate to the ANC, City Group was cognisant that the ANC did not represent all those resisting apartheid in South Africa. This critique spurred the development of City Group's 'non-sectarian' approach to solidarity and its support for a broad range of anti-apartheid tendencies; believing that the people of South Africa would determine who their legitimate representatives were through the process of struggle. This framing of the distinct role of international solidarity movements had geographical implications – their role was to prioritise exposing and disrupting the actions of those within their own society who sustained connections with apartheid South Africa. Simultaneously, they also developed their own connections with those resisting apartheid and sent material aid to the South African liberation movements. This 'non-sectarian' approach to anti-apartheid solidarity determined whom the group campaigned for and the range of political voices it listened to and took seriously. It also influenced the slogans and symbols used on the Non-Stop Picket – expanding the group's political vocabulary beyond the slogans of the ANC and its allies. In the face of hostility from some sections of the ANC's exiled leadership, in practice, City Group's 'non-sectarian' approach to solidarity work involved the development of close working relationships with members of the Pan-Africanist Congress (PAC) and the Black Consciousness Movement (of Azania) (BCMA) (Gibson 2011; Maaba 2001).

In this book, we consider social movement activism as a bundle of associated social practices and examine how elements of social movement practices become resources for individual activists (and activist peer groups) with which to negotiate specific life-course transitions. Geographers and other social scientists have tended to focus their studies of social movement activism either on the work of specific social movements (e.g. Cumbers et al. 2008; Miller 2000; Routledge 2009, 2010; Wright 2009) and/or on specific protest events (e.g. Chatterton 2006; Russell et al. 2012; Chatterton et al. 2013; Mason 2013). Our approach considers how the political engagements of anti-apartheid activists (individually and collectively) coincided with the specific demands of their lives at that point in time (c.f. Kyle et al. 2011) and how practices travel from one social movement to another over time and space. In doing this, we draw on a variety of different ways of conceptualising social practices (Reckwitz 2002; Nicolini 2012), but we draw most heavily on the form of social practice theory developed by Elizabeth Shove and her collaborators (Shove et al. 2009, 2012; Maller and Strengers 2013; Maller 2015). It is worth providing an overview of these ideas before examining how they might be applied to thinking first about solidarity, then 'growing up', through a social practice lens, and the ways in which the shared resources between these two bundles of practices might be explained.

We understand practice as 'a routinized type of behaviour' (Reckwitz 2002: 249). Practices are more than habits (Anderson 2014; Dewsbury and Bissell 2015); they consist of interdependencies between diverse elements including "forms of bodily activities, forms of mental activities, 'things' and their use, a background knowledge in the form of understanding, know-how, states of emotion and motivational knowledge" (Reckwitz 2002: 249). Practices exist both as an *entity* and a *performance*. Practices as entities exist in the form of a 'recognisable conjunction of *elements*' that can be 'drawn upon as a set of resources when doing' that activity. "It is through performance, through the immediacy of doing, that the 'pattern' provided by the practice-as-an-entity is filled out and reproduced" (Shove et al. 2012: 7). Specific practices only continue to exist as a result of the "countless recurrent enactments, each reproducing the interdependencies of which the practice is comprised" (Shove et al. 2012: 7).

Rather than appreciating the 'know-how' to undertake a practice as a personal attribute of the practitioner, a practice-based approach positions practitioners as "the *carriers* or hosts of a practice" (Shove et al. 2012: 7). In this respect, this approach to social practices takes seriously the role of materials, objects, and things in 'assembling the social' (Latour 2005). Of course, given the focus of the protest we are studying, our analysis is also aligned with recent debates about geopolitical assemblages (Dittmer 2014). In most social science, the established tendency is to follow the actors (even as, after ANT, it is recognised that the role of the actor is not limited to human beings). Social practice theory follows the practice, its constituent elements, and tracks their reconfiguration over time (Shove et al. 2012: 22).

Shove describes three elements of practices. They are *materials* ("including things, technologies, tangible physical entities, and the stuff of which objects are made"), *competences* (including "skill, know-how and technique"), and *meanings* (which includes "symbolic meanings, ideas and aspirations") (Shove et al. 2012: 14). These elements are integrated in the enactment of practices. If *competences* offer practical knowledge for participation in a practice, *meanings* offer the motivational knowledge for and the social significance of participation at any given moment (Shove et al. 2012: 23). In tracing the making and breaking of links between the elements of practices, they (Shove et al. 2012: 22) argue that it is "possible to describe and analyse change and stability without prioritising either agency or structure." In *Youth Activism and Solidarity*, in order to understand the development of practices of (anti-apartheid) solidarity (and their long-term consequences), we follow the trajectories of both the participants in the Non-Stop Picket and the (elements of the) practices that they engaged in at the time. In doing so, we recognise that there is an intersection between the lives of practices and their practitioners.

If it is important to study the conjunction of all three elements in order to understand the history of a practice, then our study is not just about the practice of solidarity, it is about the form that solidarity took in London in the 1980s. The Non-Stop Picket did not create 'solidarity' as a practice.

They were not the first to perform anti-apartheid solidarity, nor even to enact their anti-apartheid solidarity through protest outside the South African Embassy. But we argue that, in the context of the mid-1980s, they reconfigured a novel arrangement of elements to (re)constitute anti-apartheid solidarity. For practices to persist, they require a population of "more or less faithful carriers or practitioners" (Shove et al. 2012: 63). While the membership of the Anti-Apartheid Movement grew significantly in Britain throughout the 1980s, many of its members could only be relied upon to participate in occasional political demonstrations or enact their own, individualised consumer boycotts of South African goods. In contrast, through the Non-Stop Picket, City Group was able to mobilise a smaller band of committed activists towards a more intense enactment of solidarity. Thinking of the short life of hula hooping, as a practice, Shove et al. (2012: 75) argue that it may have faded "because it had no symbolic or normative anchoring: it was not strongly associated with either good or bad behaviour, with the reproduction of distinctions, or with fulfilling injunctions and obligations." In contrast, in the 1980s, opinions about apartheid were highly contested. It was difficult to be neutral on the topic. In its coming into being, the Non-Stop Picket asserted that it would continue until Nelson Mandela was released from jail and, in doing so, made the fulfillment of that objective an obligation for its supporters.

Unlike solidarity, of course, 'growing-up' is an assemblage of practices that few of us have the choice not to participate in. As Shove et al. (2012: 54) have argued, "by participating in some practices but not others, individuals locate themselves within society and in so doing simultaneously reproduce specific schemes and structures of meaning and order." Social scientists influenced by Bourdieu's (1977, 1984) theorization of taste, distinction and habitus, tend to:

> emphasize the relative positioning not of the practice but of the practitioner within the social order.... By contrast, we want to put the element of meaning at the centre of our enquiry. One method of doing so is to think about how practices are classified and how categories themselves evolve.
> (Shove et al. 2012: 54)

If, in 1950s America, "notions of youth and modernity [were] concurrently carried by practices like those of wearing jeans and leather jackets, and driving certain kinds of cars" (Shove et al. 2012: 55); we explore how an engagement with the practices of *being* non-stop against apartheid gave a group of young people the material, emotional, and symbolic resources they needed to grow up in London in the late 1980s (Goodwin et al. 2001; Brown and Yaffe 2016). The leading social movement theorist, Charles Tilly (1999: 260–262), famously argued that through their practices, social movements present displays of their Worthiness, Unity, Numbers, and Commitment (WUNC) to their members and to different publics in order to amplify and legitimise their political message. Through its persistent, continuous protest, the Non-Stop Picket made itself an irritant (to the police, to the diplomats inside the Embassy, to the British and South African governments,

Figure 2.1 City Group leaflet advertising the launch of the Non-Stop Picket, 19 April 1986. Source: Gavin Brown.

and to the mainstream Anti-Apartheid Movement). The picketers endured frequent harassment (petty and otherwise) from the Metropolitan Police, abuse from supporters of the South African regime, and the worst the British weather could throw at them. They claimed these cumulative hardships and micro-aggressions, and wore them as a badge of honour – a marker of their worthiness and commitment. This book not only pays attention to the spectacular, theatrical moments of protest when City Group displayed their WUNC most effectively. It also attends to the power of the small interactions between picketers. It considers how these interactions and practices made larger contentious acts possible, but also contributed to the ways in which youthful protesters 'grew up', and adult campaigners negotiated the multiple pressures affecting their lives (Figure 2.1).

Launching the Non-Stop Picket

The Non-Stop Picket of the South African Embassy started on 19 April 1986. A few days before the launch, on 15 April, the US bombed Tripoli in Libya, and many core activists feared that the protests against the American military action might detract from their plans to launch the Non-Stop Picket. On the day, City Group members and their supporters marched from Camden Town Hall, near Kings Cross, to the South African Embassy. Their march was led by a group of children[29] carrying a patchwork banner that had been hand sewn by Norma Kitson.

> There was a very, very colourful banner at the front which was held by children and it was a remarkable sight. There were, to my view, about 1,500 people on that march. And we marched down to Trafalgar Square and there was this remarkable sight at Trafalgar Square because there seemed to be hundreds and hundreds of people waiting for us to arrive there to start the Non-Stop Picket. So, it was a remarkable event.[30]

> On the first day of the Non-Stop Picket we had a march from Camden Town Hall to Trafalgar Square and Carol was the Chief Steward and I was the Chief Steward's Assistant.... It was a really amazing day. We had planned the demo but then the US bombed Libya the same week. So there was a big demo at the US embassy on the same day, and lots of people who might have wanted to go to one or the other were, obviously, a bit torn. But, it didn't matter. We had a great demo and then lots of people came down afterwards from that and we carried on.[31]

In preparation for the arrival of the demonstration, a small number of City Group activists were delegated to transport and set up the equipment needed for the Picket and its launch rally. Beyond this technical role, one of their key tasks was to claim and secure the group's protest site in front of the Embassy gates.

As well as claiming space and putting the Picket's minimal infrastructure of banners, placards, and supplies of publicity material in place, the planning for the launch of the Non-Stop Picket had also involved preparing a number of contingencies to ensure that the first night and the follow few days were covered by volunteers. As Non-Stop Picket activist, James, explained, for him, there were mixed emotions as the launch rally came to an end:

> It was filled with the exuberance of doing what we'd said we'd do and what we'd planned to do for a while. And then, arriving there, there were the inevitable peaks and troughs that were going to take place (which I hadn't thought about beforehand)… And I remember, as nightfall came and everyone had thrown their big energies into the day, we had contingencies in place and people lined up to do the evening and night shifts, and those of us who had been busy all day moving stuff around took breaks. It was a bit strange to be going away – there had been so many people before and now there were so few. Are we going to be able to do it?[32]

Carol Brickley, one of the more experienced activists involved in launching the Picket, was perhaps more realistic in her assessment of the challenge City Group had set itself. While many of the younger picketers remember the excitement of the day, in her capacity as the Group's Convenor, she offered a political assessment of the role of the Non-Stop Picket, linking it clearly to events in South Africa:

> I thought it was something that you take on, but you don't project yourself to the end of it. You don't say I am doing this because it's possible, because frankly it didn't seem altogether possible. But the value of it was in the doing of it rather than in the end of it. Especially given what was going on in South Africa. I mean, it's not that you decide that sort of thing in isolation.[33]

The Non-Stop Picket did not just appear out of nowhere. It built upon City Group's experience of regularly picketing the South African Embassy (at least weekly, but often more frequently) since early 1982. During the previous four years, City Group had fought to defend the right to protest outside the South African Embassy and was determined to maintain that right. Their call for a Non-Stop Picket of the Embassy responded to growing anti-apartheid militancy inside South Africa and captured the imagination of young people in Britain who wanted to play an active role in opposing apartheid. While the potential conflict with the US Embassy demonstration on 19 April 1986 understandably made City Group activists fearful that their plans to launch the Non-Stop Picket might be undermined; in the end, the coincidence probably worked in their favour by attracting additional young activists along on the day who were open to City Group's anti-imperialist analysis of the need to oppose apartheid. This book tells their story.

Notes

1 O'Neill, S. (1989), "Demo of a thousand days," *City Limits* (12–19 January), London Voice Ltd: London. p. 7.

2   During the Spring Adjournment debate on Wednesday 25 May 1988 (HC Deb 25 May 1988 vol. 134 cc351-91), John Carlisle, the Conservative MP for Luton North, described the Non-Stop Picket as "the continuing demonstration outside the South African Embassy is becoming rather more than a nuisance." He demanded that the Metropolitan Police be given more powers to remove the anti-apartheid protest. His speech came a year after City Group had successfully fought against an attempt to ban them from protesting outside the Embassy (see Chapter 5). It gives some indication of quite how much the Picket's continuous presence outside the South African Embassy for the previous two years had riled the representatives of the apartheid state inside the building (and their supporters in Britain).

In hindsight, the MP's description of the picketers as "a bunch of left-wing political extremists," "rent-a-mob," and a "motley crew" is perhaps not so wide of the mark. Indeed, as often happened when they were attacked in such ways, the picketers took these insults, turned them on their head, and made them a badge of honour. The phrase "We are a motley crew...." was adapted into some of the Picket's regular chants in the weeks that followed Carlisle's attack.

3   Raw footage of a Free Steve Kitson Campaign picket of the South African Embassy can be seen at: www.itnsource.com/en/shotlist/RTV/1982/01/11/RT-V110182002/?s=%22south+africa+house%22+protest&st=0&pn=1 (Accessed 2 August 2016).

4   After the unbanning of the SACP, David Kitson (1991) published a stinging 'anti-revisionist' critique of the Party's political programme and praxis, questioning whether it was 'really communist'.

5   The Justice for Kitson Campaign was launched at a meeting of trade unionists and anti-apartheid activists on 21 May 1988 at Camden Town Hall, London. K. Fernand (1988), 'Justice for David Kitson', *Non-Stop Against Apartheid*, 28 (June), p. 7; Justice *for Kitson Campaign* pamphlet (1988), London.

6   In fact, his mother's ANC unit in London had, until this point, advised Steven against any direct political involvement, in order that he might continue to be allowed to visit his father in gaol each year (Kitson 1987: 248)

7   The Revolutionary Community Group (RCG) has its origins in the 'Revolutionary Opposition' faction within the International Socialists in the early 1970s around Roy Tearse. The faction itself soon split after their expulsion from IS. Under David Yaffe's leadership, the RCG was formed in 1974 and began a period of intense theoretical work about the nature of the crisis of capitalism and the role of the labour aristocracy. See: www.revolutionarycommunist.org/gallery1/1490-25-years-old-frfi (Accessed 2 August 2016).

8   LBC/IRN, "Kitson family on aunt's murder in South Africa" 15 January 1982. http://bufvc.ac.uk/tvandradio/lbc/index.php/segment/0000300173012 (Accessed 2 August 2016).

9   In this context, in its initial existence, the group was briefly known as City of London Anti-Apartheid Branch.

10 In a document sent to the ANC's National Executive Committee from the UK Regional Political Committee entitled 'Memo from the RPC further to the recommendation for the expulsion from membership of the ANC of Norma and David Kitson' (3 February 1987), ANC officials quote extensively from Norma's autobiography (Kitson 1987) to demonstrate how she had breached internal security and publicised internal divisions within the ANC (especially tensions between white and African members). Mayibuye Archives: MCH-02 ANC London Box 14, F: untitled bright blue filing cabinet folder.

11 It was not until the ANC's consultative conference at Morogoro in Tanzania in April 1969 (five years after David Kitson's arrest) that people of all racial backgrounds were welcomed into membership of the ANC (Ellis 2013; Macmillan 2013).

12 Joe Slovo had been active in the Communist Party of South Africa (later SACP) from the 1940s, becoming one of its key theoreticians. He was a leading member

of Umkhonto we Sizwe in exile, serving as its Chief of Staff until April 1987. In 1984, he became General Secretary of the SACP, and in 1985, he became the first white member of the ANC's national executive.

13 Interview with Ann Elliot.
14 City of London Anti-Apartheid Group (1984), 'Press statement – 10 June 1984'. Copy held in City Group office papers and amongst Steve Kitson's personal papers.
15 City of London Anti-Apartheid Group (1984), 'Press statement – 10 June 1984'.
16 Statement of Witness by Commander George Howlett of Cannon Row Police Station, Metropolitan Police, 6 July 1984. Copy in Steve Kitson's personal papers.
17 Figures for the exact number of arrestees are difficult to find. Andy Higginbottom's Secretary's report to the City Group AGM on 16 February 1986 claimed there had been 169 arrests during the campaign. However, it should be remembered that a number of people were arrested multiple times, hence there will have been more arrests than arrestees.
18 'London Borough of Camden powers of demonstration', legal opinion by Stephen Sedley QC, prepared for the London Borough of Camden, 6 July 1984. Copy in Steve Kitson's personal papers.
19 Bodleian Library, Oxford, MSS AAM 503: Statement issued by the Anti-Apartheid Movement Executive Committee, 10 July 1984.
20 City Group was described as a 'hooligan element' of the Anti-Apartheid Movement in a briefing note (reference TXS 027/334/1) by the Protocol Department of the Foreign and Commonwealth Office, 11 May 1984, in preparation for the visit to the UK by President Botha, obtained under the Freedom of Information Act and supplied to the authors by the journalist Solomon Hughes.
21 *Fight Racism! Fight Imperialism!* March 1985, issue 47, p. 1.
22 Two members of the Anti-Apartheid Movement Archives Committee, Christabel Gurney (former Editor of Anti-Apartheid News) and David Kenvyn (former Chair of London Committee of the AAM) both made this point in informal conversations with Gavin Brown during our research.
23 Bodleian Library, Oxford, MSS AAM 21: *The Anti-Apartheid Movement and City AA: A Statement by the AAM Executive Committee*, 1 December 1985.
24 This was Gavin's first contact with City Group, as a 15-year-old. It put City Group on his radar, but he did not become a regular participant in their events for another nine months or so.
25 www.itnsource.com/en/shotlist/ITN/1985/10/19/AS191085008/ (Accessed 2 August 2016).
26 This chant was a reference of the ongoing controversy regarding South African intelligence operations out of the Embassy targeting exiled South African campaigners and British solidarity activists (Trewhela 1995; Bell with Ntsebeza 2003).
27 A Higginbottom, 'Report on Defence Campaigns', paper submitted to the City of London Anti-Apartheid Group Annual General Meeting, held at County Hall, London, 16 February 1986.
28 *Non-Stop Picket against Apartheid outside the South African Embassy in Trafalgar Square, 1986* [video]. RCG/FRFI. www.youtube.com/watch?v=0cD0tJvzSQ4&list=PLOY0wSGz5-LrunonDBiifSY7kv1antVcJ&index=1&feature=plpp_video (Accessed 19 June 2013).
29 This group of children included Helen Yaffe, then aged 8, and her sister.
30 Interview with David Yaffe.
31 Interview with Nicki.
32 Interview with James Godfrey.
33 Interview with Carol Brickley.

3   Becoming non-stop

Compared to the imposing edifice of South Africa House, the infrastructure of the Non-Stop Picket was flimsy and ephemeral – a banner and a few boxes – yet picketers succeeded in making their protest highly visible, day and night. The success and longevity of the Non-Stop Picket benefited from the location of South Africa House – the South African Embassy occupies the whole of the east side of Trafalgar Square. At the time, Trafalgar Square was not only a major tourist destination, but it was also the main hub of London's night-bus network. The square was, to some extent, busy at all hours of the day and night. Twenty-four hours a day, people walked by the Picket – often on their way between the West End and one of the mainline rail stations in the area. This constant flow of people gave the Picket a public to engage with and a source of new recruits to refresh its ranks. As one of the police officers who occasionally guarded the Embassy during that period noted:

> It almost became normal and part of Trafalgar Square, like Nelson's Column. It was a totally accepted part of London life.[1]

Another benefit of its location was that Morley's Hill, the stretch of pavement in front of the Embassy on the outer edge of the square, was approximately 5 m wide, making it one of the widest pavements in London. However hard they might try, this made it difficult for the police to claim that the protest routinely obstructed the flow of pedestrians using the thoroughfare. If the South African Embassy had been located on a quiet leafy square in Mayfair or Belgravia (like the US Embassy), it would have had little visibility and would have been considerably harder to sustain. In so many ways, the location of the Embassy and the monumental architecture of South Africa House worked to the advantage of the anti-apartheid protesters:

> I mean we couldn't have asked for better. You know, we had this big white building that was so symbolic of the apartheid regime, and it provided such a perfect focus point for protest, with huge visibility.[2]

Through its presence and its practices, the Non-Stop Picket laid claim to and used the pavement as public space (Mitchell 2003; Mitchell and Staeheli 2005) – a place in the city where diverse people could stop and engage in political debate. While City Group actively sought to make the Picket appealing, welcoming, and hospitable – a place where people felt they could stop and participate – at the same time, picketers quickly came to feel very territorial about 'their' pavement.

This chapter introduces some of the young people (and others) who were attracted to the Non-Stop Picket by its continuous presence in Trafalgar Square. It examines how these young people first came into contact with the Picket and what it was about the politics and practices of the Non-Stop Picket that enticed them to stay and get involved. Some came from highly privileged backgrounds; others were living homeless on the streets. Some joined out of curiosity; others out of existing political commitments. Despite their diverse social backgrounds, many participants remember that they got involved at a time when they were feeling 'lost' or as if something was 'missing' in their lives. We suggest that this is quite a common experience in the transition from adolescence and young adulthood (even if these young people found an uncommon route to self-exploration) (Brown and Yaffe 2016)

The picket infrastructure

For such a long-standing protest, the physical infrastructure of the Non-Stop Picket was minimal and unsubstantial. Using very little permanent equipment, the Picket made its presence felt. The fabric of the Non-Stop Picket grew over time, but started out as little more than a hand-sewn banner proclaiming the protest's purpose and a few boxes of equipment and provisions that were essential to the effectiveness of its campaigning. Over time, the Picket's infrastructure grew slightly, as the Group won legal and political battles with the police about how permanent their base could be. But, from the beginning of the Non-Stop Picket, City Group drew on its experience of four years of previous protests outside the Embassy (see Chapter 1) to ensure that its flimsy structures and the activists who sustained it had the greatest visual impact. Very quickly, the Picket developed a distinctive appearance, using the black, green, and gold colours of the African National Congress (ANC) on banners, hand-drawn placards worn by protesters, and other equipment to visually declare its politics and *raison d'etre* (Figure 3.1).

Many picketers, like Andy Privett, remember the Non-Stop Picket as being "vibrant, inviting and colourful," acknowledging that it was deliberately like this in order "to attract attention."[3] When asked to describe the physical appearance of the Non-Stop Picket, picketers frequently focussed on the banner. Quite literally, it was central to the fabric of the Picket:

Figure 3.1 Fight Racism! Fight Imperialism! banner in front of South African House, photographer unknown. Source: City Group.

> There was a large banner held up by two fixed poles. It was usually colourful and often had a portrait of Nelson Mandela. The banner was changed from time to time and was usually skilfully made by Annie.[4]

In the early days of the Non-Stop Picket, the wooden poles holding up the banner had to be held at all times. As Claus remembered:

> In the summer of 1986 it felt like an honour to hold the banner, after the first winter it became more and more a punishment.[5]

In Trafalgar Square, at the time, it seemed like the prevailing wind blew eastwards, down the Mall, across the square, and hit the Non-Stop Picket with its full force. Keeping the banner vertical became physically tiring after a while; and in strong winds, it was an exhausting challenge. While the banners were updated frequently in order to remain topical and to publicise forthcoming large protests, being perpetually exposed to the elements, they had a relatively short lifespan before they needed to be washed and repaired.

In the early months of the Picket, the banner was often the only equipment that was evident immediately in front of the embassy gates. In an attempt to make protesting as difficult as possible, the Metropolitan Police prevented the picketers from sitting down or placing *any* equipment on the ground in front of the embassy gates (see Chapter 5). Over time, as the Picket became more confident in its presence (and staying power), the police were forced to ease some of their more punitive restrictions on the Picket, and its infrastructure developed: to make picketing a little less physically demanding, to help the Picket function more effectively, look more appealing to the public, and communicate its message clearly:

> I remember [Claus] from Holland, he had a nifty idea at some stage to use a car tyre and put a pole in it and fill it with concrete, set it, and then we did not have to hold the banner any more so that was quite an invention. It took years before we actually got round to doing that, but that really worked well.[6]

In addition to inventing a self-supporting banner mechanism, at around the same time, the Picket acquired three sturdy storage boxes with wooden lids painted in black, green, and gold. These were lined up beneath the banner in the centre of the protest and used to store the Group's equipment and publicity material. They replaced the battered (and frequently soggy) cardboard boxes of equipment that had previously been stored beneath the tree several metres up the pavement, near the corner of Duncannon Street. One additional benefit of these new boxes was that they were strong enough for weary picketers to sit down on and take a short rest (if the police officers on duty were not being too officious).

Other, more permanent equipment was also accumulated and put to use:

> There were usually several placards on the ground (when there were many picketers, some would hold them or wear them). Sometimes they gave facts; sometimes they focused on a current event or campaign. Often they had general slogans like: "Black Majority Rule," "One Person, One Vote," "One Settler, One Bullet" etc. These changed very often.[7]

> I remember the bucket obviously for the donations, and obviously placards as well. Not always the placards because, obviously, if it was too windy they blew all over the place, but I think we used to have some stones to weight them down as well sometimes.[8]

As Simon notes, although the picketers were creative in finding ways to communicate their cause and attract new supporters, the experience of being outside in all weathers sometimes limited what equipment could be used (and how long it stayed looking appealing). When we interviewed them, many former picketers acknowledged that the Non-Stop Picket frequently ended up looking messy, untidy, and a little shabby. Some also remembered that they spent a lot of time on the picket worrying about its appearance and obsessively trying to keep it tidy. In many ways, this was always a losing battle, and by the end of the Non-Stop Picket, the legacy of four years of continuous protest had become sedimented in place, leaving its mark on the pavement in front of South Africa House:

> It was messy. I think that's what [Claus] tried to change because, where people were standing there the whole time, just the pavement itself was black and dirty, people's coffee and all these things spilt or they'd leave food, and so it was a … So generally it was a bit, yeah its appearance was messy and scruffy, and lots of us on it were messy and scruffy as well.[9]

The Picket's practices

Those messy, scruffy picketers stood on that pavement in Trafalgar Square for a reason – to demonstrate their opposition to apartheid and to encourage others to get involved in anti-apartheid campaigning. Time spent on the Picket was time spent being routinely engaged in a range of different campaigning practices. Although the Picket was reproduced through a repertoire of habitual practices, which activities were being performed at any given moment tended to be shaped by who was there, how many people were there, the time of day, and a variety of other factors:

> You could have two men and his dog there one day and then the next day there might be a whole heap of people. So, and because it was such a diverse group of people, some days might be more, you might see more people selling papers; some days you might see people who were more confident on the megaphone, so it would be more noisy. But you would try to always make sure you had two people to hold the banner.[10]

There was a clear hierarchy of activities. If (before Claus invented the self-supporting banner pole) the main priority was to keep the banner on display, the moment there were more than two people present, petitioning passing members of the public became the main priority. Much of the material stored on the picket was associated with this practice – there were printed petition forms that people could sign, there were pens for signing, and there were clipboards for people to lean on. Indeed, for most former picketers, clipboards will forever be known as 'petition boards'.

When their numbers permitted, two, three, or half a dozen protesters would stand in a line, parallel to the embassy walls, and about a third of the way in from the kerb across the pavement. Resting on their forearms would be a petition board. As members of the public walked by, they would be asked to "sign the petition for the release of Nelson Mandela." Once someone had stopped to sign the petition, a number of associated practices might be enacted (although these could depend on the skill and confidence of the petitioner). In addition to their signature, they would be asked to make a donation. Sometime donations were kept in a small plastic pouch clipped to the petition board, but in later years, they were deposited into the large donations bucket. Having signed the petition, the passer-by would be given a leaflet for the Group's next demonstration or major protest. If they had made a donation of more than a few pence, they might be given a copy of the Group's newsletter *Non-Stop News* (later renamed *Non-Stop Against Apartheid*[11]). If the petitioner was a supporter of the Revolutionary Communist Group, they might also try to sell a copy of their newspaper *Fight Racism! Fight Imperialism!* a practice that, at times, was opposed by some picketers who were not affiliated with that political organisation.[12] Those people who signed the petition and seemed particularly supportive of the Picket would be encouraged to stop and join in for a while, or to come back when they could.

Signatures on the petition were a valuable campaigning asset (especially when they could be counted in the hundreds of thousands) – they helped the Group to assert the worthiness of their cause. This measure of public support also helped to protect the Picket in the face of intensified political policing. However, the practice of petitioning was crucial to the continuation of the Picket in other ways – donations collected helped fund the Group's campaigning, and conversations initiated across a petition board helped recruit many future picket stalwarts to the cause. Even so, petitioning was contingent on the weather, and great care was taken to protect the petitions and propaganda from the rain.

> I remember times when it would be lashing down with rain and all the placards would get wet and you would be desperately trying to get all of the leaflets in the boxes, and not just in the boxes, but ensuring that leaflet A was on top of leaflets A, and leaflets B did not mixed up with the petition sheets, and make sure that the signatures didn't get smudged, because they were important things, the petitions. OK, you couldn't petition when it was raining, but you wanted to be in a position to start petitioning the moment the rain stopped and you couldn't do that if the things were wet.[13]

One of the other defining features of the Non-Stop Picket was its noise. The Picket could be quiet, if the number of protesters present was low, or if it was the middle of the night, but very often it was alive with sound (Kanngieser 2012; Revill 2016). During the day, picketers would use one of the Group's megaphones to make speeches about apartheid or lead others in chanting. Singing was also a core practice on the picket. Picketers sang songs created out of the struggle against apartheid in South Africa (Gilbert 2007); but other songs followed, improvised on the picket to celebrate its work. Singing was so central to the life of the Picket that the lyrics of these South African freedom songs were printed on song sheets for distribution on the protest to encourage all picketers to join in. Often, songs were taught, phonetically, line by line. This entailed a certain degree of trust, as picketers learned to

Figure 3.2 City Group Singers perform on the Non-Stop Picket. Source: Photographer: Jon Kempster.

sing songs in Southern African languages they (usually) did not understand (Figure 3.2). Retired Police Inspector David Lee, remembers:

> They were noisy and not always aggressive. The chanting (in African style) was very catchy and had a great beat. I have seen much worse.[14]

City Group also had a choir, City Group Singers, with a loose and constantly shifting membership. Very often, the Singers would try out new songs and, having rehearsed them, would teach them to other picketers. As Ken Hughes, a key member of City Group Singers and a notable singer remembered:

> City Group singers were born out of I suppose having learned all these songs on the picket and having then started going out to various other rallies and marches and, you know, sometimes talks in universities or whatever. It was quite a nice thing actually to be able to go along there, introduce a couple of South African freedom songs, and if you like the City Group singers grew out of a group of people who were keen to learn a lot of the songs, more than just the ones that were sung regularly on the picket and use them on big rallies, on big events, because it was undoubtedly a good focal point for attracting people onto our contingent. And it was also a good morale booster really when you're in a sort of drudgery of a march from Hyde Park to Trafalgar Square, you know, walking through places that you knew people who didn't give a toss about what you were doing. It kind of kept me going really, yeah.[15]

As Ken noted, the Singers performed at many rallies and events organised by City Group, but they also played an important role in fundraising for the Picket and Southern African liberation movements through busking sessions and sales of their cassette, *Freedom Songs*,[16] released in November 1987.

Singing played an important role in the political culture of the Non-Stop Picket in many other ways (Wood et al. 2007). The practice of collective singing brought picketers together (Mattern 1998). In this way, it helped activate the Picket when energy was flagging, providing focus and impetus to keep going with the political work of petitioning, fundraising and engaging passing members of the public in dialogue. If the Picket was being threatened by police intervention or neo-fascist attack, choral singing could help cohere the Picket as a collective so that it was better able to defend itself.

Even though many protesters sang with more enthusiasm than skill, singing songs from the struggle against apartheid was itself a form of solidarity, a means of spreading a political message, and symbolically sharing a culture of resistance with those opposing racist rule in South Africa (Gilbert 2007). One of the easier songs for new picketers to learn was *Bayangena*, which had some simple dance steps associated with it. As they sang, picketers would edge forwards, verse by verse, getting closer to the embassy walls, testing how close they could get before the police on duty ordered them back. This is a good illustration of how song became integrated into other aspects of the Non-Stop Picket: energizing picketers, keeping them warm, and as a means to assert the Picket's disrespect for the authority of apartheid's representatives inside the Embassy.

Geographies of the Picket

The geography of the Non-Stop Picket was broader than the few square metres of pavement occupied by anti-apartheid protesters outside the South African Embassy. Certainly that is the most obvious way to think of *where* the Non-Stop Picket was, but the Non-Stop Picket was produced and sustained through activities that took place in a constellation of locations beyond that stretch of pavement. There was the network of locations in which City Group activists protested against apartheid; there was the Group's office (a couple of miles east of Trafalgar Square) and the various venues in which the Group held meetings to plan and organise its activism; and there were various more mundane sites around the West End that sustained the Picket (Feigenbaum et al. 2013; Brown et al. 2017).

The Non-Stop Picket became part of the geography of the West End. There were a variety of locations in the vicinity where picketers rested, socialised, and were socialised into being 'non-stop against apartheid'. There were the various cafés and pubs where picketers went to eat, to warm up, to use the toilet, or to relax after a shift. Some of these venues were constants throughout the four years of the Non-Stop Picket, and some (particularly several of the pubs) only welcomed the custom of picketers for a short period. Which venues people used also varied throughout the day and throughout the week. During the day, most days, there was a steady flow of picketers in and out of the Breadline Cafe on Duncannon Street. Picketers would pop in there for takeaway teas and coffees to warm themselves up on cold, wet days. Often, groups of the younger picketers would hang out there gossiping, especially on weekends. Breadline had the advantage of being cheap, and the staff were (mostly) friendly and sympathetic to the Picket. A pricier option was the Café in the Crypt of St Martin-in-the-Fields Church, also on Duncannon Street. The Crypt had the advantage of toilets; and, while many picketers would eat there or chat over coffee, many others would sneak in just to use the toilet when they needed to.

For those picketers who socialised with their fellow activists, many evenings were spent drinking in pubs around the area. Over the four years of the Picket, many different pubs within a short walk of Trafalgar Square were popular with City Group activists. Although it is fair to say that City Group, especially when drinking *en masse*, were not always popular in those pubs, and it was not uncommon for the Group to be barred from pubs in the area.

The options of where to eat, to pick up a hot drink, or use the toilet were more limited for those on the overnight shift. For them, coffee runs meant a trip to the all-night café near Leicester Square. When their options were limited, picketers often faced ethical dilemmas about where to patronise. As Francis recalls:

> I'm ashamed to admit that we would often go to McShit [McDonalds] after the night shift as it was the only place open. I can't imagine ever going there now.[17]

The Non-Stop Picket did not exist in isolation from the urban spaces around it. The practices of being a non-stop picketer extended beyond those that were conducted directly in front of the Embassy in a protest space. Being non-stop against apartheid frequently entailed a broader set of social practices that took place across Central London. As the Non-Stop Picket inserted itself into the quotidian geographies of the West End of London, more and more future picketers came into contact with the protest. In turn, as people with anti-apartheid sympathies became non-stop picketers, they found ways to incorporate time spent on the picket into their habitual use of the city. Inevitably, in the process, their relationship to the geography of Central London changed.

Who was non-stop?

This book predominantly focusses on the experiences of the young people who stood on the Non-Stop Picket, but the protest attracted a very wide range of participants. Many were young, but the Picket was supported by people of all ages, including some stalwart pensioners. Most of the protesters were white, but the Non-Stop Picket was more successful at mobilising black and Asian protesters than many events organised by the national Anti-Apartheid Movement (Fieldhouse 2005: 478; Williams 2015). Many former picketers evocatively captured the diversity of people who came together in a shared endeavour on that pavement:

> And there would be many multi-coloured placards and really multi-coloured people. I mean there were like punk rockers, there were hippies, there were mods-I just remember it was really diverse and people of every age and ethnicity and ...[18]
>
> All sorts of people came, young and old and different ethnicities, men and women and girls and boys, people who worked in respectable jobs, people who did not work and people whose occupation might well have been outside the law – all sorts.[19]

Here, we introduce some of the young people (and others) who were attracted to the Non-Stop Picket and helped sustain its continuous presence in Trafalgar Square.

The Non-Stop Picket did not arise out of nowhere. As we outlined in Chapter 2, from its inception in 1982, City Group brought together friends of the Kitson family alongside Carol Brickley and her comrades in the Revolutionary Communist Group (RCG). Some of the pensioners on the Picket had worked with Norma and David in the Communist Party of Great Britain in the 1950s. From its early days, City Group also attracted current and former classmates of both Steve and Amandla Kitson. Through the first non-stop picket in 1982 and the Group's subsequent activities, several of them became core members of the Group's committee by the start of the Non-Stop Picket in 1986. At its inception, then, the Non-Stop Picket contained a number of established and committed activists of various generations (including people who had first become politically active as young people through City Group's activities earlier in the 1980s).

> I was communist, committed to fighting for a new kind of society, a socialist society. I was anti-imperialist, totally opposed to British intervention overseas, especially in underdeveloped countries and before the picket obviously I was opposed to British intervention in relation to South Africa.[20]
>
> How would I describe myself? I think I was about 21, and I'd moved from Edinburgh to London. Yes, so that was my personal circumstances. I was involved in politics, had been for about three or four years previously. I had quite a young involvement. I was a member at that time of the Revolutionary Communist Group.[21]

In addition to the Kitson family, over time, the Non-Stop Picket also attracted the support and participation of a small number of other South African exiles (and visitors) who had direct experience of anti-apartheid organising in that country. Their presence added credibility to the protest and strengthened other participants' sense of a connection to the struggle in South Africa.

By far the largest group of people to become involved in the Non-Stop Picket (and the majority of those we interviewed through our research) were young people who were becoming interested in politics, often for the first time. Even so, they did not constitute a uniform group. There were school students, university students, and recent graduates, as well as teenagers and those in their early twenties who were working or unemployed. Most had grown up in Britain, but there was a significant minority who had lived elsewhere in the world for most of their lives before they got involved in the protest. One or two specifically relocated to London in order to join the Non-Stop Picket. Whatever their background, many of them were (as Eleanor Cave, one of their number, described herself) "Young, idealistic, opinionated, [and] loudmouthed!"[22]

Time and again, former picketers who had joined when they were aged about sixteen described how their commitment to the Non-Stop Picket was the culmination of a growing interest in politics over the preceding few years. For many, the Miners' Strike of 1984–1985 had been a defining moment in their politicisation (and it is worth remembering that for those who grew up in the southeast London suburbs, the Kent coalfield was not that far from their childhood homes) (Kelliher 2017). For Sharon, a 16-year-old junior secretary from Orpington who joined the Picket with Georgina, her best friend from school, this was certainly the case.

> I was young and pretty naïve about politics but had supported the miners' strike and the printers' strike at Wapping. I didn't grow up in a political family. We didn't discuss politics much and the only newspapers we had in the house were tabloids (i.e. The Sun and the News of the World). I was beginning to realize that there was injustice in the world and wanted to be part of something that could change things.[23]

Like many other young people in Britain at the time, other picketers, like Helen Marsden from Camberley in Surrey, had been politicised by the fear of nuclear annihilation as the Cold War heated up again in the early 1980s. Several young activists joined the Non-Stop Picket having first become active in the Youth Campaign for Nuclear Disarmament (YCND) and the broader anti-nuclear movement (McKay 2004; Smith and Worley 2014):

> Well, I guess I was beginning my political journey. I think really my journey started when I was about nine and I joined the RSPCA, and I think that's really where it came from because I realized that people

were cruel to animals, and I thought that was wrong ... And then probably when I was about 15/16 I got involved with CND. So, because we were already paranoid in the 80s about nuclear attacks and how imminent they were and everybody watched that programme *Threads* about a nuclear bomb landing on Sheffield. So we were all living in fear I think then. So I then went to places like Greenham [and] Aldermaston.[24]

Grace Livingstone was part of a group of friends attending a comprehensive school in North London. They had begun to become politicised together about international geopolitics at school and initially attended the Non-Stop Picket together as a group:

We were at Haverstock ... So basically my friends and I had become a little bit politicized already, because we'd heard about Nicaragua and we were trying to like twin our school with that, and so on. But not involved in political activity apart from maybe trying to twin our school with that.[25]

Mark, who went on to become the Picket Organiser later in the Picket's existence, also joined as a sixth-form student. He was already passionate about politics but had been disappointed by the response he had received from the Anti-Apartheid Movement when he asked about ways of getting involved in campaigning. For him, the Non-Stop Picket provided an opportunity to do something about an international issue that mattered to him:

I was a school kid. I was in the sixth form at a school in South London and I'd been interested in politics and I used to spend my sort of Christmas or birthday present money joining campaign groups and things like that. And I'd contacted the Anti-Apartheid Movement and they sent me about 20 sort of Pillars of Apartheid posters with companies on them-that was it. So when I asked for more information and to get involved I just got 20 identical folded A3 posters and nothing else.[26]

Alongside these young people who had grown up in London and its suburbs, the Picket also attracted young people who had recently arrived in London and were looking for somewhere to belong (which is not to suggest that they – or, at least, some of them – were not also serious about politics). Deirdre's story is typical of this group. Later, when her sister also moved to London from Ireland, she too connected with the Non-Stop Picket. This is how Deirdre remembers herself at the time:

Isolated I suppose. I moved to London at 18 for the summer, then came back, and then came back just before my 19th birthday. So I was working in Croydon, I knew very few people in London.[27]

Not all of the young people who joined the protest were teenagers. Many, like Sally, were young adults in their early twenties:

> I was 22 years old and doing a degree foundation course at Thames Poly. I was not politically active but was politically aware. [...] I had been given a leaflet about the Picket whilst in a queue for a gig. In the beginning I went to the Picket on Saturdays with my sister Jane and some friends and their children. Every week all of us really looked forward to going and on the train up we would get really excited. As we got off the train and walked to the embassy our pace quickened especially if we heard singing or chanting. We were always made to feel really welcome.[28]

For Sally, going to the Picket was a collective, social event she shared with members of her family and friends. The excitement of attending the Picket and the excitement of going to the West End seem entwined in her recollections (Skelton 2013b). Sally first engaged with the Picket as she was embarking on a new phase in her life and a return to studying as a mature student.

Many of the adults who joined City Group after the Non-Stop Picket started did so as they found themselves in a significant life transition (McLeod and Thomson 2009; Hörschelmann 2011). Penelope had recently graduated, as a mature student, from the University of Warwick, but in many ways, her life was still in flux, and she had not yet found a niche for herself in London that she was content with. Her description of her life at the time gives a rich insight into aspects of London's social geography during the Thatcher era that is often overlooked:

> [I was] newly graduated and unable or disinclined to get a swanky job in Thatcher's high unemployment Britain – being a manager would have been against my principles, I was adrift socially and employment-wise, dating people from *City Limits* adverts to no avail, and doing casual and voluntary work, reading for the blind, occasionally writing for and distributing 'Battleaxe' – a women's liberation mag for women in the labour movement. I had applied to become a bus driver, driving 9s, 11s and 19s to pay off my mortgage and was waiting to hear. I had just completed a women's carpentry course. I was more feminist than socialist.[29]

For two of the men who joined the Picket in their late twenties and early thirties, their engagement with the protest coincided with the breakdown of their marriages. Mike, a former soldier who dabbled in stand-up comedy, stumbled across the Picket by chance in the autumn of 1986 but quickly threw himself into the Group's activities. Geopolitical events at the time had led him to start reappraising his politics and their role in his life:

> My personal life was up the shitter, I mean I'd just been divorced and I had my heart broken. And it's like the old cliché of going off to join the French Foreign Legion; I thought this would be the end of me. I thought if I threw myself in this struggle, because the struggle was big enough to take anything I've got, I don't even know if I'll make a dent, and that's okay. That was where I was, my personal bullshit was there, emotionally I was there.[30]

In contrast to Mike, Simon had more of a track record of political activity on the Left. He was active in his trade union and was in the process of leaving the Labour Party in protest of its drift to the Right under Kinnock's leadership. Although he had previously been aware of the Non-Stop Picket, he did not commit to the protest until early 1989 when he attended the celebration of the Picket's 1000th day (see Chapter 4). Initially, he joined with a friend, but she did not sustain the same level of involvement:

> My friend Lorna lost interest after about three months I think but I carried on, because at that point my marriage was falling apart and I think like a lot of people who gravitated to the Picket, it was something, you'd lost something and the Picket was filling that void, if you like, and meeting up with all the people that I did.[31]

Simon was not the only picketer to get involved in the Picket during a time when their life was difficult. Many participants remembered that they got involved at a time when they felt as if something was 'missing' in their lives. In many ways, this sense of not fitting in is quite a common experience for young people in the transition from adolescence to young adulthood (Valentine 2003; Horton and Kraftl 2006; Worth 2009). In this regard, the young picketers were not necessarily particularly extraordinary (even if they found a less common route for their self-exploration and self-realization). As we have already noted, for some, this process of self-discovery was associated with having the space to explore what London could offer them and having time and space for autonomous decision making away from their family homes. Daniel, who joined the Picket when he was 15, is typical in this regard:

> So by the time I became involved with the picket I used to often come up [to central London] in the evenings – just get on the train straight from school and come up to the picket and just hang out there. The picket was used very much by a lot of people I think, you know, as much for its social dimensions as other things and there were a lot of people who had similar kind of strange families or kind of funny home situations that it felt more comfortable on the picket, that's kind of how I felt.[32]

Reflecting on his time on the Picket during its final two years, Mark Bearn offered some fascinating observations about the political effects of the casual and habitual mixing of people *with a purpose* on the Non-Stop Picket. He recognised that, for many participants, however deep their political commitment to the anti-apartheid cause was, their connection to the Picket also fulfilled other personal and social needs. The Non-Stop Picket connected with many different strands of everyday life in London simultaneously; but it also transformed how its participants experienced life in the city:

> Like most others I found a solidarity in the picket I have rarely found since. The eclecticism of the membership, and the diversity of encounters one had with members of the public and casual visitors, was a political education in itself. Any prejudices I might have had about age, gender, sexuality, personality, politics, crumbled pretty swiftly-I have never felt such a sense of unity. Unity that was all the more effective for being casual and everyday at times. Obviously many people involved-no doubt including me-had motives that were selfish in some way: I was often struck by how many people seemed to be there for friendship, community, even love. Yet the fact that we were there in support of a cause that benefited very few of us personally, that we supported for moral and ethical reasons as much as political ones, seemed to me a noble thing in the end.[33]

In early 1989, the *Mail on Sunday's You* magazine published a feature on the Non-Stop Picket.[34] This article offered fascinating insight into how the Picket and its participants were seen, by those outside the Left, at the time. The article tells a story shaped for the newspaper's conservative readership, but the dozen or so picketers that it quoted and described were broadly representative of the range of people who sustained the Picket in its later years. That the Picket offered a means for "self-confessed alcoholic" homeless men to protest alongside Quaker doctors, middle-class young women looking for something more than the suburbs could offer them, and "disarmingly cheery" Scottish revolutionaries is to its credit. The article positions the Picket firmly within the social and political geography of London in the Thatcher era and highlights how the South African Embassy's location in Trafalgar Square made the Picket visible and accessible to a wide range of people in London at the time.

Making contact

As should already be apparent, future picketers came into contact with the Non-Stop Picket in a variety of ways. Some found out about the Picket through contact with City Group activists at other political events. Many came along for the first time with a friend or relative, either out of mutual curiosity or because that friend was already involved with the Picket and had encouraged them to join too. But, a surprisingly large number found the Picket through a chance encounter (Wilson 2012; Merrifield 2013; Askins 2015) – literally stumbling across it as they pursued other interests in Central London and finding it enticing.

From the first day of the Non-Stop Picket onwards, people joined the protest because they had been given a leaflet by (or had had a conversation about it with) existing members of City Group at other protests. Throughout the duration of the Picket, City Group would regularly send contingents on other progressive demonstrations and protests in London as a show of mutual solidarity, but also as a means of spreading word of their own activities. The City Group archives are full of invitations to participate from other groups and campaigns.[35] This proved particularly effective in attracting new supporters, especially when a major event was being planned.

If City Group sent out organised contingents to mobilise support at other protests, individual picketers were also persistent in encouraging their friends, relatives, and colleagues to support the Non-Stop Picket. We interviewed several former picketers whose siblings had also been involved in anti-apartheid campaigning. Some picketers admitted that they first attended the Picket because they wanted to emulate an older sibling:

> My elder sister was involved and I wanted to be like her. Then I came back to UK from two years away in Canada and started to live with her and as her friends were politically active so I also became so too.[36]

Two other interviews, with people who had attended the same school for a while, demonstrate how an involvement with the Picket could ripple through a peer group or friendship network (Bosco 2007; Bowlby 2011; Bunnell et al. 2012):

> In the sixth form we had a sort of afternoon off and some people that had been there, they had (during the summer holidays) they'd come across the picket and started joining it. So then I think it was Wednesday afternoons when we had a time when we were meant to be doing work and then we used to go together.[37]

> In my school in south London I met a boy who was active on the picket. I didn't have much in common with the other kids in school so we became friends and he introduced me to the picket.[38]

Having joined the Picket through (and with) school friends, Mark soon found that within a few weeks, he was the only member of his group who continued to use the time set about by their school for 'volunteering in the community' as an opportunity to protest against apartheid. However, in Sigþrúður's case and others, he would continue to find friends at the school who were prepared to get involved. Sigþrúður, whose academic parents were on sabbatical in London at the time, had few friends in the city and was another person whose connection to the Picket helped them overcome social isolation and a sense of being adrift in the capital.

The Picket's constant presence in Trafalgar Square meant that many thousands of people passed it every day. Many of those people would have taken little notice of it, a few were actively hostile to it, but others were inspired by what they saw. Time and again, former picketers recalled first encountering the protest by chance as they went about their business in the area. Kathy, a Labour Party activist who worked in the West End, found the Picket while taking a walk after work one day:

> At the time I was working in Covent Garden, for a solicitors in Covent Garden, and used to go to down to Trafalgar Square during lunch breaks and everything and one evening I passed the South African Embassy and there was a group there. I think that was, I think the autumn time of [19]86 and the Non-Stop Picket was there.[39]

Other people encountered the Picket on their way to or from the National Gallery, the theatre, or another leisure activity in Trafalgar Square or the wider West End. Tunde, who had spent part of his life in Nigeria, discovered the demonstration in this way. For him, this chance encounter provided him with the opportunity he had been looking for to take action against apartheid:

> So it was an opportunity for me when I was walking sometime around May 1986 towards the Trafalgar Square and I saw the demonstration so I just joined straight up. To me it was like the perfect thing for me to do, you know, and I'd been looking forward to doing something like that.[40]

Betta, from Brazil, first saw the Picket as a tourist. She was so inspired that she returned to look for them when she returned to live in London a year later:

> In 1987 I spent a week in London and passing by Trafalgar Square as a tourist I saw the picket and was really impressed by it … then in 1988 I moved to London to improve my English and decided to see if that amazing group of people was still there and it was. So I immediately got involved.[41]

Like so many other newcomers to London, Betta saw the Picket as a chance to connect with other young Londoners who she believed shared her outlook on the world. The politics of the Picket were important to her, but so too was the opportunity it afforded her to make herself part of the city she was temporarily living in.

On weekend overnight shifts, the Picket was provided with a hot drink and sustaining food by the Simon Community (a charity promoting self-help for and by the homeless) and the 'soup runs' of other charities supporting the street homeless in Central London (Johnsen et al. 2005). Julie initially came into contact with the Picket as a volunteer with one of these charities, Bondway. She remembers that she spent a lot of time on the Picket during its first year:

> I was homeless at the time so [it] was a great support and also worked on the soup run initially, which used to stop there. I was more a supporter than a hard-core activist, but I did my bit.[42]

Another person who found the Picket through the experience of being homeless on the streets of the West End was Chris:

> I was a teenage runaway back in 1986. When I arrived in London, like many I headed up to the West End of London and found St Martin's church. With nowhere to sleep I learnt about the demonstrations and stayed a while. For a naive 17-year-old from the Midlands it strangely gave a safe environment ... I think I stayed nearly a week, learning about London, where to get a safe sleep, and ironically learning about Mandela ... After moving to a hostel nearby, I used to come down and sit on the steps of St Martin's.[43]

Through its constant presence in Trafalgar Square, the Non-Stop Picket provided nearby homeless youth with a place to hang out, engage with people, and get informal advice, where they were not necessarily defined by their homelessness.

It would, however, be false to suggest that all those who went on to commit their efforts to the Non-Stop Picket had a happy and inspiring first encounter with it (Valentine 2008). In the final months of the Picket, Francis spent a considerable amount of time on the protest, often staying there overnight for several nights in a row. Nevertheless, although sympathetic to the cause, a simple misreading of the function of part of the Picket's infrastructure left him feeling attacked and put off by the 'strange' people who shouted at him:

> Once, before I joined the picket, I threw some litter in the [donations] bin, thinking it was for that purpose. Two girls who were sitting on the boxes shouted "fascist!" I tried to walk over to explain what happened when two young men walked me away saying "just go, just go!" I thought they were very strange![44]

Francis's story suggests that, although chance encounters on the streets of Central London were important in bringing the Non-Stop Picket to the attention of many potential supporters, those encounters alone were not necessarily enough to convert a sympathetic member of the public into a new picketer. The quality of that encounter was also crucial (Matejskova and Leitner 2011; Wilson 2012, 2016). In Francis's case, it was a long-standing commitment to anti-racism and a need to get involved in something during a difficult time in his life that drove him back, despite the tricky first encounter.

Getting involved

City Group developed a political culture and method of organising that sought to be welcoming and inclusive. The Picket fitted into the wider geography and rhythms of life in Central London at the time, and that enabled its supporters to engage with the protest in a number of ways. Although some people who came into contact with the Picket went on to make an on-going commitment to spending regular time there, others found ways to weave a more occasional engagement with the protest into their lives. The Non-Stop Picket connected with the daily and weekly cycles of the city in other ways too. Debbie told us:

> I used to visit the Non-Stop Picket from time to time, mainly accompanying a friend who did a regular early morning slot before she went to work! It's all pretty vague, but I do remember having a sense of belonging to a family.[45]

If Debbie and her friend attended the Picket in the early mornings on the way to work, other people fitted their attendance around work in other ways. In 1989, Fiona was 19 and working as a nanny in London:

> When I could, I visited the National Gallery as much as possible, and the protest that was always outside South Africa House made me curious. Once I knew what it was about, I had to be involved … For 18 months I was part of the picket in 1989/90. I picketed during the day on my days off and would stay until the last tube home mostly.[46]

For Graham, who then worked "as motorcycle courier, or on market stalls," visits to the Picket were fitted in more around the rhythms of his social life than work commitments. He offers a fascinating insight into how the Non-Stop Picket served multiple functions for its different constituencies and fitted into many different circuits of the West End (Brown 2008, 2009):

> I live at the Elephant, and often walked home from clubbing in the West End. I knew about the picket, and was happy to join it, especially in the small hours when I could get some fresh air before walking home. I never questioned why there were so many gay men there, until someone told me years later it was a cruising site. I had naively assumed that the gay community was merely in solidarity with the South Africans, and never thought it further. I spent many hours chatting away, not realising what was going on, happy to be supporting the struggle, before leaving for home.[47]

Beyond spending time on the picket line, some people in the West End found other means of demonstrating their support for the anti-apartheid cause. For example, Liz Myers (and others) remembered how staff from a local vegetarian restaurant would donate unsold food to the Picket at the end of business each day:

> I wonder who remembers looking forward to the left overs from Cranks in Covent Garden coming down? Cheers if it was almost-warm pizzas-groans if it was cheese scones.[48]

As we discuss in Chapter 4, the Non-Stop Picket was organised through a weekly rota, which served to formalize individuals' involvement in the protest. Many of those who did make a regular commitment to the Picket's rota and went on to become core City Group activists for a period of time recalled that it was the warmth of the reception that they received on first approaching the Picket that enticed them to return. Although she could not remember the precise details of whom she spoke to, Irene remembered a sense of the Picket being "very welcoming" and attracting a "different group of people" that she felt comfortable around and with whom she felt an affinity. Others could remember specific picketers that they engaged with – very often experienced picketers who (although still in their twenties) had been involved with City Group since before the Non-Stop Picket began – who welcomed them, made them feel at home, and encouraged them to stay and play a part in the protest. A handful of names were mentioned repeatedly as being particularly effective in this role, including Richard Roques, City Group's Treasurer. For Georgina, Lorna Reid played that role (but other names cropped up too):

> I really remember the essence of Lorna, very sort of friendly, sort of like very warm and as I say enough to make us feel able to go back again.[49]

For Mike Burgess, having stayed up all night chatting with picketers on an overnight shift the first time he stumbled across the protests, it was a lengthy conversation with Steve Kitson the following morning that encouraged him to commit to becoming non-stop against apartheid:

> They get me in conversation and I end up standing there for all night chit-chatting about apartheid. Then in the morning Steven Kitson arrives for his Saturday morning shift, which he used to do every other Saturday morning. And he told me his story. And he told me David Kitson's story, and I was sold.[50]

Mike was not alone in committing to the Non-Stop Picket immediately on first encountering it – the Picket had an infectious quality like that (Anderson 2014). However, more commonly, people were enticed back by the urgency of the cause and the friendliness of the established picketers and found that, over time, they were spending more and more time there, more often, and started to participate in a wider range of the Group's activities. Attending the Non-Stop Picket became habitual (Bissell 2015):

> We would come past the picket a couple of times and speak to people there and gradually, I can't really remember exactly, but I suppose just seeing some of the same people there week after week, you know, you'd kind of get to know faces and say hello to people and then kind of hang out a bit more and just try and find out what was going on. And so I think it was probably for me quite a gradual thing.[51]

Although many people found the Picket enticing and exciting, it could also be "scary and alienating" (Penelope Reynolds) for those who tried to get involved without knowing anyone else there already. The friendly welcome offered by established picketers was a crucial practice – the sooner newcomers felt a connection to other protesters, the more likely they were to return. Similarly, offering new arrivals a task to do on the Picket – frequently petitioning the public or distributing leaflets – allowed them to feel important, included, and as if they were making a positive contribution. There was one further practice that played an important role in converting sympathisers into active members of the protest, and that was asking all those who joined the Picket to sign into the 'picket book'. This logbook became a record of how many had participated on any given shift (sometimes crucial in tracing witnesses to arrests), but it also captured the contact details of all participants. Where people gave an address, they would be added onto City Group's mailing list. When they gave a phone number, they could expect a phone call from the Group encouraging them to get involved in future activities. In April 1986, Cat Weiner was a student in Oxford. She attended the Non-Stop Picket on the day it was launched and returned a few times after that, whenever she could make it to London, without feeling that she was necessarily committing to the protest. However, a phone call from the City Group office helped convince her to become more involved:

> [At] some point not that long afterwards I got a phone call to say would I come to a meeting on a Friday, and a few weeks in, I must have been coming to the a picket for a couple of weekends and I'd be coming down on a Saturday and someone said can you come down on a Friday? Because I did overnight shifts on Saturdays, mainly sort of Saturday afternoon onwards, so I started coming down I think fortnightly on a Friday for meetings and got more involved from that point really and had a regular shift on the rota.[52]

As should be apparent from many of the quotes presented here, the Non-Stop Picket attracted a very diverse range of people to its cause. Some people got drawn into the protest and made regular commitments to the Picket's rota. In some cases, these commitments lasted for years. Other people were regular supporters but never formally pledged their time in that way – they came when they could. Yet, other participants had a short, intense engagement with the Picket and then moved on to other places and other interests. Whatever their level of commitment, many people's first contact with the Picket was incidental to their presence in the West End of London for other purposes. They found ways to incorporate time on the Non-Stop Picket into the specific rhythms of *their* time in Central London. It was in this way that disparate individuals felt they could make a useful contribution.

Forging new connections

The Non-Stop Picket existed to protest against apartheid and to build solidarity with those resisting apartheid in Southern Africa. City Group believed that one of its central tasks was to mobilise a mass solidarity movement in Britain that could force the British government and British companies to sever their ties with apartheid South Africa. Four months into the existence of the Non-Stop Picket, they published an assessment of the impact of that on-going protest:

> Many new people have become involved for the first time in the struggle against apartheid by joining the Non-Stop Picket. There are thousands more throughout London and Britain who want to take action. We have to let them know about the Non-Stop Picket. City Group is organising for that. We need you to help.[53]

The article went on to list six actions that supporters could take to mobilise new people to take action against apartheid with and through the Non-Stop Picket. Many of these actions, such as placing posters about the Picket in radical bookshops and community centres, allude to aspects of London's urban geography that are now seriously depleted. The article studiously avoided calling on activists to fly-post posters on the streets (which would have been illegal) – although this was a common practice at the time and a staple means of publicising protests (Pimlott 2011).

Through its central location in Trafalgar Square and the excitement it generated through its vibrant culture of protest and confrontation with the representatives of apartheid in Britain, the Non-Stop Picket attracted a broad and diverse group of activists from the UK and beyond. City Group developed a suite of practices through which to engage the public with their cause and encourage new people to get involved. These practices were effective in attracting a diverse group of anti-apartheid protesters to participate in the Picket (although they occasionally backfired and could be as alienating and off-putting as they were enticing).

For the most part though, picketers tried to make the space of their protest as welcoming as possible. The minimal and fragile infrastructure of the Picket was used to best effect to publicise its anti-apartheid message. However, the realities of running a continual protest with little equipment, outdoors in all weathers, meant that the Picket often looked shabby and less attractive than was hoped. Some picketers invested a disproportionate amount of their time on the Picket tidying up and cleaning around the protest. Even so, after four years, the Picket's continuous presence left its mark on those few square metres of the east side of Trafalgar Square, as spilled coffee and discarded cigarette ash became sedimented on the flagstones where the picketers had stood.

Being on a wide pavement in one of the busiest public thoroughfares in London significantly improved accessibility for the Non-Stop Picket and assisted in building the levels of support it enjoyed, and its longevity. Many people, from many different walks of life, first encountered the Non-Stop Picket by chance as they participated in other aspects of the everyday life of Central London. Because of where it was located, the Picket was found by commuters on their way to and from work, clubbers on their way home after a night out dancing, homeless youth waiting to be fed by a soup run late at night, or young Quakers on a weekend retreat. All these people and more found their way to the Picket. Some came because they were already committed to the cause; others because they were curious; and others still because they were bored, lonely, or feeling adrift in the city. Frequently, they stayed simply because someone took the time to ask them to get involved and come back.

The Picket provided 'uncommon ground' through which friendship networks developed that crossed boundaries of nationality, ethnicity, age, class, and social difference (Chatterton 2006; Harris and Valentine 2017). Brought together by their common opposition to apartheid, young picketers protested alongside, and socialised with, people from very different backgrounds to their own. Beyond the space that the Picket occupied on the pavement in front of the embassy gates, the picketers adopted other spaces around the areas as part of their territory – places where they went for a snack, for the toilet, or to warm up during picket shifts; places where they held their meetings and planned their campaigns; and the places where they socialised with others who were *non-stop against apartheid*.

Notes

1 Interview with a retired woman Police Constable from Special Patrol Group. This officer served in the Special Patrol Group in Central London from 1980 until 1986. She had family connections in Rhodesia and South Africa and admits that coloured her opinion of anti-apartheid protesters at the time. She medically retired from the police in the 1990s and is now a magistrate in the West of England.
2 Interview with Deirdre Healy.

3 Interview with Andrew Privett.
4 Interview with Francis Squire.
5 Interview with Claus.
6 Interview with Andre Schott.
7 Interview with Francis Squire.
8 Interview with Simon Murray.
9 Interview with Mark Farmaner.
10 Interview with Helen Marsden.
11 A frequent, but irregular newsletter was published throughout the Non-Stop Picket. Twenty-two issues of *Non-Stop News* were published in the first 18 months of the Non-Stop Picket, with the first issue of the renamed *Non-Stop Against Apartheid* (No. 23) being published in October 1987.
12 These tensions recurred (with a shifting cast of complainants) throughout the Non-Stop Picket and City Group's longer existence. At times, the tensions ran right to the heart of the Group – on 5 July 1986, Norma Kitson convened a meeting of 'non-aligned' activists at her house to discuss their concerns about the RCG and to develop the political leadership of City Group activists who were not supporters of the RCG. (Copies of the minutes of this meeting are held with the papers from the City Group office).
13 Interview with Vincent.
14 Email correspondence with former Metropolitan Police Inspector David Lee. David Lee joined the Metropolitan Police in 1968, having served in the Merchant Navy. He was promoted to Inspector in 1980 and joined the Area Support Unit based at Cannon Row (which became the Territorial Support Group). He retired in 1998.
15 Interview with Ken Bodden.
16 City Group Singers (1987), *Freedom Songs from the Release Mandela Non-Stop Picket, 1986–1987*, London: Aluta. The cassette has been digitised and is available at https://soundcloud.com/gavinbrown/were-here-till-mandelas-free (Accessed 2 August 2016).
17 Interview with Francis Squire.
18 Interview with Susan Yaffe.
19 Interview with Trevor Rayne.
20 Interview with David Yaffe.
21 Interview with Lorna Reid.
22 Interview with Eleanor Cave.
23 Interview with Sharon Chisholm.
24 Interview with Helen Marsden.
25 Interview with Grace Livingstone.
26 Interview with Mark Farmaner.
27 Interview with Deirdre Healy.
28 Email correspondence with Sally O'Donnell.
29 Interview with Penelope Reynolds.
30 Interview with Mike Burgess.
31 Interview with Simon Murray.
32 Interview with Daniel Jewesbury.
33 Email correspondence with Mark Bearn.
34 K Fletcher (1989), "Nelson's Column," *Mail on Sunday* (*You* magazine), undated (possibly 8 January), pp. 62–64.
35 For example, amongst other correspondence, the City Group archives contain: Letter from the Free the Guildford Four campaign inviting City Group to sponsor their campaign, 25 June 1986; Letter from the Committee for the Defence of Democratic Rights in Turkey inviting participation in their event, 27 October

1986; Letter from Lambeth Women's Rights Committee inviting City Group Singers to perform at Women Internationally day of activities, 3 March 1987.
36 Interview with Jacky Sutton.
37 Interview with Mark Farmaner.
38 Interview with Sigþrúður Gunnarsdóttir.
39 Interview with Kathy Fernand.
40 Interview with Tunde Forrest.
41 Interview with Betta Garavaldi.
42 Email correspondence with Julie.
43 Email correspondence with Chris.
44 Interview with Francis Squire.
45 Email correspondence with Debbie.
46 Email correspondence with Fiona Brownlie.
47 Email correspondence with Graham Neale.
48 Email correspondence with Liz Myers.
49 Interview with Georgina Lansbury.
50 Interview with Mike Burgess.
51 Interview with Daniel Jewesbury.
52 Interview with Cat Weiner.
53 City of London Anti-Apartheid Group (1986) 'Build the Picket!' *Non-Stop News*, No. 9 (8 August), p. 4.

4 Being non-stop against apartheid

Life on (and around) the Non-Stop Picket

The Non-Stop Picket had a distinct geography, as we argued in the previous chapter. The Picket occupied a certain (shifting) space in Trafalgar Square as the main site for City Group's anti-apartheid protests. But the Picket was also part of a broader constellation of sites where picketers organised their campaigns, reproduced their capacity to protest, and socialised (with) each other as a protest collective. Through its constant presence in Trafalgar Square, the protest became aligned with the distinct geographies of the West End's many different social constituencies – people differently positioned within Central London's division of labour, tourists, clubbers, and the street homeless, amongst many others. Each of these groups experienced the Picket and the urban spaces around it through distinct, but synchronic, rhythms of life – they passed the Picket at different times, with different degrees of urgency, as they went to and fro between the different sites and practices that structured their lives (Lefebvre 2004; Pantzar and Shove 2010; Edensor 2012). In the previous chapter, we primarily described the geography of the Non-Stop Picket and analysed the lived experience of the spaces that constituted it. Now, we turn our attention to the temporalities (Hoffman 2011; Sharma 2014) – the lived experience of time – on and around the Non-Stop Picket. To spend time on the Non-Stop Picket was to experience time in a very particular way. For nearly four years, it *was* non-stop. We ground our analysis in an attention to the different practices that (notionally) structured the functioning of the Picket and then examine the different ways in which these may have varied across the day and night, as they responded to the changing rhythms of the city around them.

Sustaining a 'non-stop' protest around an 'urgent' global issue required non-stop commitment from core activists that was frequently hard to sustain. In addition to considering the temporalities of life on the Non-Stop Picket, this chapter considers how that pace of activity fits with the experience of youth and the transition to adulthood. At times, the collective energy of picketers flagged and the Picket struggled to keep going. To maintain momentum the Picket was structured around particular weekly rituals and an annual calendar of events. The Picket found ways of celebrating its longevity that served to recognise the commitment of existing activists and recruit new participants. In considering the way time passed and was marked on the Picket, this chapter examines the different rhythms of the protest – its daily, weekly and annual cycles. It does so by examining some of the special events that celebrated the Picket's anniversaries and key achievements, such as its 1000th day.

The Chief Steward and the rota

The Non-Stop Picket was "ruthlessly well-organised"[1] and run according to a series of rules that had been agreed and established in practice during City Group's earlier protests outside South Africa House. Unlike the functional anarchism of some of other long-term protests in 1980s Britain, such as the Greenham Common Women's Peace Camp or the Faslane Peace Camp (Roseneil 2000), the Non-Stop Picket was highly organised, and City Group operated through a hierarchical leadership structure. Political leadership was provided by the Group's Convenor (Carol Brickley), Deputy Convenor (Norma Kitson), and Secretary (a post held by various activists over time, with Andy Higginbottom as the incumbent for most of the duration of the Non-Stop Picket[2]). In addition to these three officers, there was a committee (comprised of a dozen or more people) that met weekly to plan and organise the Group's campaigning. At different points in time, various committee officers convened sub-groups of a broader layer of volunteers to assist them in their duties.

The Picket was organised through a weekly rota of 3- and 6-hour shifts. Individual supporters were encouraged to commit to one or more regular shifts (preferably weekly, but as regularly as they could manage). A volunteer Picket Organiser worked to ensure that shifts on the Picket's rota were filled, in advance, with adequate personnel each week. Although the Picket Organisers worked hard to ensure every shift on the rota was covered, there were inevitably occasions when people failed to turn up, and the previous shift was left stranded:

> You felt responsible that you had to do it. ... I think once or twice, generally, yeah there were moments when I've been there on my own. But not for very long, because normally someone off the street would stop and talk to you, you know, or someone would eventually turn up.[3]

On each shift, one picketer acted as the Picket's Chief Steward. For the most part, this responsibility was taken by an experienced and trusted (but not necessarily older) picketer; but a competent and committed activist could quickly find themselves in this role (especially on those shifts that were harder to cover on the rota). The role of the Steward was, first and foremost, a political one. As an internal 'user guide' to the Picket, circulated in the final few months of the Picket in early 1990, stated:

The rules and policies of the Picket are established at democratic meetings of City of London Anti-Apartheid Group. These rules and policies obtain despite the disposition of the Picket at any given time. It is the responsibility of the Chief Steward to defend what City Group stands for. There is a political duty to ensure that the message gets through. The Chief Steward is the 'custodian' of the Picket's political message.[4]

The rules referred to here were simple, few in number, and stood throughout the duration of the Picket. These rules served to ensure the political integrity of the protest and the safety of its participants. They banned the use drugs and alcohol on the picket and stated that no one under the influence of either was allowed to be present there. This was not a moral position (except, perhaps, for Norma Kitson), but one motivated by the security of the Picket, to reduce the potential for drunk picketers to provoke unnecessary arrests. Participants were encouraged to fully participate in the political work of the Picket and to ensure that it was tidy and presentable. The rules expressly forbade picketers from engaging in conversations with the police (or responding to taunts and abuse from racists).

Richard recalled in more detail the different tasks that comprised the Chief Steward's role and how they were used to create continuity between picket shifts:

> You would arrive and you would have to talk to the [previous] chief steward about anything that had happened. There was a book that you wrote things down in. You wrote down, if you were the chief steward, that you'd arrived and you have to sign things, and there was a book that people were allowed to write in which was just about their experiences. And that's if you're a chief steward … and then you would ask people to sign the petition, you would organise chants on the megaphone and you would, if there were a lot of people there, try and talk to new people, try and get their details. Try and build the Picket, try and start some singing going, and try and get a sense of community, particularly if there were new people there who you wanted to encourage to return, you didn't want it to be boring for them, you wanted things to happen.[5]

The Chief Steward wore an armband tied out of red ribbon to identify themselves to other picketers and to the police on duty. City Group asserted to the police that they were only to communicate with picketers via the Chief Steward. They were also entrusted with carrying a small shoulder bag that contained a number of items that were crucial to the role. The 'steward's bag' contained the steward's notebook (mentioned by Richard) in which, for each shift, the Steward kept contemporaneous notes of any incidents on the picket – conflict with the police, arrests, troublesome picketers who breached the rules, or an uplifting visit by a passing celebrity. In the event of arrests, these notes could be valuable evidence for the Group's lawyers.[6] The steward also carried a camera with which to document similar incidents. Finally, they were entrusted to safeguard the Picket's finances and would carry any paper banknotes that had been donated (until they could be taken away to the Group's office for banking). In this way, the Steward was enrolled in an assemblage of 'disobedient objects' that served as technologies for the effective reproduction of the protest (Davies 2012; Flood and Grindon 2014).

Important though these practices were, the Steward's role was not just bureaucratic and procedural. It was a political role – the Steward oversaw the political work of the Picket and the clarity of its message, but an effective steward also understood that this required activating and energizing the picketers on the shift:

> You'd come on, you'd take the bag, you'd take the megaphone. I think the thing that I always remember feeling is the sheer … what I really loved is, I remember sometimes Lorna having the megaphone and as people came up, because she'd be chanting, always chanting and then she'd get you joined in, you know, she might say your name as you're coming up… She had this way of just getting your energy levels up before you'd even started, and I really liked that. I always used to think it was really important to keep the singing and the chanting going.[7]

As Andre acknowledged, although certain authority was vested in the Chief Steward on any given shift, the impulse to provide political leadership on the picket was held in common by a wider layer of core activists:

> We were incredibly well organised, and there were a lot of people who were being driven to show leadership and so maybe not the whole, [not] everybody on the Picket at any one time would be that driving force, but probably at least half or a third of them would be. So yes, shall we do some petitioning now, let's get the petition boards out, and then it was about trying to raise money, trying to see how much we could possibly raise. Like I said, the megaphone, trying to make a speech or two. Some of the people would sing and it's amazing how even, it could be just a small group and we had the confidence to sing some South African song.[8]

City Group's officers, committees, and cadre of committed activists provided a governance infrastructure (Feigenbaum et al. 2013) to steer the actions of the Non-Stop Picket (Simone 2004). The rota formed the central organisational infrastructure of the protest and facilitated its continuous presence for nearly four years. Beyond this, the Picket's rules and the hierarchical leadership of each picket shift (although contested by some who passed through, and even stayed on, the Picket) provided a continuity in how the protest functioned from one shift to another. Despite this continuity, in practice, the experience of time spent on the Picket (and the ways in which those rules and procedures were applied) could vary considerably in any 24-hour period.

Daily rhythms of the Picket

> I remember that it was a bit like an amoeba in that it would change shape over a day, certainly over the months of the year and over the years. So sometimes it would be, I know this, very small, and sometimes it would stretch out along the pavement, and then there were the big rallies and big occasions of course. And so its shape changed considerably[9].

As Ann remembers, although the Non-Stop Picket was a constant presence in Trafalgar Square, the actual space that it occupied pulsed continually, expanding and shrinking across the day and across the week. This oscillation was affected by more than just the time of day, or the day of the week; it changed according to the weather, the mood of the protesters who were present at any given moment, and external political factors such as events in South Africa.

> In the pouring rain in mid-winter we all looked bleak and bedraggled but a friend might turn up and it would be as if the sun had come out and romance was sometimes in the air.[10]

Despite the rules and procedures that were supposed to produce consistent conduct on the picket line, many former picketers remembered that in reality, there was no such thing as a 'typical shift'. If there was typicality, then, as Lorna acknowledged, it was that the political work of the Non-Stop Picket became quotidian and unexceptional:

> I think it was different at different times of the day. If you were there in the morning obviously it was very quiet. You saw London wake up, and it would gradually get, the footfall would increase as the day went on. And then numbers on the picket itself would increase accordingly. A typical shift? Probably mostly uneventful. It was about maintaining a presence. It wasn't all high jinks and excitement; it was very much a commitment to having the presence outside the Embassy.[11]

Time spent on the picket line was scored by a series of overlapping rhythms of different duration. The volume of people attending the Picket and the pace at which they undertook their political work increased in accordance with the rhythms of the city around it. During the daytime and early evening, picketers concentrated on interacting with as many members of the passing public as they could; "speaking through the megaphone, leafleting, stopping people to sign the petition."[12] In contrast, the night shifts were (for Simone) shaped more by the commitment to maintain a constant presence and felt "more like guard duty."[13] The time of day not only affected what was likely to be occurring on and around the Non-Stop Picket, but it could also determine some of the practices that could occur there. For example, following a protracted dispute with the police and local council over 'noise pollution' in the early months of the Picket's existence, picketers were prohibited from using their megaphones before 9 o'clock in the morning, acknowledging that there was a small residential population in the vicinity (see Chapter 5).

The time of day, the point in the week, and the season could all affect the mood of the people passing the Picket in Trafalgar Square. This, of course, had an impact on how the picketers interacted with them – there was often little point approaching people rushing to work in the morning; but, although talkative, people who had spent the night out enjoying themselves could be unpredictable and pose their own challenges:

> Completely varied, I mean if I was doing a 7 until 10 shift [in the morning] at some point, I mean that was just dead, there was nothing there. People rushing by on their way to work, no one stopping or anything like that. On a Saturday night it was bonkers and it was always a very interesting one. So at that time in Trafalgar Square all the night buses went from there so the entire West End and people from the Southbank and other places would all come out of pubs. So it would be busy with people going to clubs and then people leaving pubs and then the clubs would empty out and I think you had Heaven around the corner and when that emptied out, you know, there was just so many people that had been taking drugs and stuff and dressed up in strange costumes and things like that and being very, very happy.[14]

Despite these regular patterns that could be anticipated, the rhythms of the Non-Stop Picket were also affected by other factors, originating well beyond Trafalgar Square or the West End of London, as Ann recalled:

> A typical shift, I don't think you could say there was anything typical about the shifts there, because it was such a fluctuating situation depending on the time of year, the time of day and the political environment, what was happening with South Africa and in London, so I don't think there was anything typical about any of the shifts at all.[15]

Although there could be lots of excitement to be had on the Non-Stop Picket, sometimes the hours dragged by slowly, and picketing could be boring. That was not always a bad thing for picketers. Sometimes periods of boredom led to moments of creativity (Toohey 2011). Time committed to the Picket could move tediously slowly, and creative forms of activism were generated to break or alleviate these moments of boredom (Anderson 2004).

> I was lucky as well that I initially did a shift with somebody who loves singing too, so I got to learn South African songs, that wouldn't

necessarily happen on all shifts. We often did sing. I think myself and Tim used to sing a lot even if it was just the two of us.[16]

Singing and dancing helped pass the time and helped picketers to keep warm through long winter shifts. When there were few passers-by around to engage in political discussion, or ask to sign the petition for Mandela's release, picketers sometimes made up songs and raps about the anti-apartheid struggle, or celebrating their own experiences of being on the picket. On hearing the news of the assassination of President Samora Machel of Mozambique (by South African forces) (Westad 2005: 362), the picketers spontaneously devised a simple, haunting song in his memory.[17] There were also raps celebrating the women on the Picket and pillorying some of the more aggressive and notorious officers from Cannon Row Police Station responsible for policing the Picket. Although not all of these songs might have been developed on the picket in dull moments, they are indicative of how humour and other creative expressions were used there to (re)produce forms of geopolitical knowledge, cohere the picketers as a collective, and frame their opposition to apartheid (Hart 2007; Dittmer 2013; Dodds and Kirby 2013).

This creative activism was *as* entangled with the rhythms and emotions of sustaining the Non-Stop Picket as a long-term activist project, *as much as* it stemmed from a desire to communicate a political cause in accessible forms. Creative expressions of solidarity both helped attract new participants and strengthen the bonds and shared commitments between existing picketers– they were a demonstration of 'WUNC', demonstrating the Worthiness, Unity, Numbers, and Commitment of the picketers (Tilly 1999). They had a performative effect in gelling the Group as a solidarity collective (Brown and Yaffe 2014). Over time, songs and chants were repeated so frequently that they are amongst the aspects of the Picket that participants remember most clearly.

Trafalgar Square nights

At night, the Non-Stop Picket fitted into the geography and temporal rhythms of Central London in quite different ways. Jacky's description of her memories of nights spent there is evocative of the different waves of activity that punctuated the life of the Picket in the depth of the night. Her memories of overnight shifts on the Picket are shaped not only by the limited facilities that picketers could draw on, nor the unusual mix of people they could expect to encounter on the streets at that hour, but they are deeply embodied and etched by the cold in her bones and the sour smells of the early morning city (Stoller 1997; Longhurst 2001):

> I cycled from Stoke Newington and got to Trafalgar Square at about 10.30 p.m. I'd get a coffee at the American Diner place in Leicester Square and then go and put my bike round behind St Martins. I'd go and check it periodically but it was an old bike. Then I'd go and stand

on the picket and light up a cigarette to keep myself awake. The first couple of hours was always fun because the people would leave the pubs and bars and come and heckle and argue. Central London at night is great, especially with the soup kitchen vans passing by and all the religious nuts hanging around. We'd often have mad or homeless people stopping by for a chat, to insult us or to cadge cigarettes off. When the McDonald's shut (I think around 1am) we'd have to go to the British Council area for the loo so I didn't go that often. I'd wait until [I think] 5.30 a.m. when the McDonald's opened and then go to warm my hands in the hot water in the basin and have a foul coffee. Dawn was the worse time because the temperature dropped a couple of degrees – I don't remember the Picket in the summer, I just remember the bone chilling cold seeping up from the pavement and the smell of sour milk, vomit and car exhaust that heralded the new day.[18]

In the late 1980s, Trafalgar Square remained active all night long. The anti-apartheid protesters always found people to share (part of) the night with them. Overnight shifts on the Non-Stop Picket were qualitatively different from daytime shifts, and some picketers chose to undertake them because of this. With fewer people around passing the Picket, political work was less outward-looking (Massey 2005; Darling 2009) and conducted at a slower pace:

> At night we'd sometimes sit or lounge on the floor and just chat. We weren't supposed to do that, but it did get more chilled at night.[19]
>
> Overnight shifts were generally quieter.... I think you became closer to the other Picketers on these shifts.[20]

Although officially the Picket's rules applied as much to night shifts as they did during the day, many former picketers acknowledged that (depending on the personality of the person acting as Chief Steward) the rules were frequently applied less rigidly and with more nuance in the middle of the night. For example, the strict rule against talking to the police officers on duty softened on night shifts. Picketers would still be cautious about what they said to (and in front of) the police, but many were more relaxed about unexpectedly silly interactions when they occurred:

> But then [the police] would mellow out after midnight. Maybe someone had a guitar there. I do remember one evening where somebody had a guitar and decided to sing all of the seventeen verses of, or however many it is, of American Pie, with one of the coppers there, one of the Cannon Row coppers who obviously knew all the words, and he was singing along and you'd think these are very, very bizarre moments.[21]

The police, too, tended to be less rigid in their attitudes to regulating the Picket at night. For picketers who were used to frequent arrests for petty infringements during the day, this change of attitude often helped them see particular officers as individuals and identify those who were (apparently) more sympathetic to the anti-apartheid cause.

> Just a few of you standing there facing the cops – there was more banter and baiting than during the day, which changed your opinion of the cops. Some were right bastards and others were quite sympathetic, helpful even.[22]

Another rule that was (sometimes) applied less strictly on night shifts was the one that barred the participation of picketers when they were under the influence of alcohol. Someone who was obviously drunk would probably be sent away, as their attendance could still present a risk to themselves or others. Picketers would try to ensure they were not seen drinking on the picket, but that did not mean that some would not seek opportunities for a drink in the middle of the night, as Gary admitted:

> We used to drink quite a lot there actually ... Not necessarily on the picket, but it was secret – I mean you'd have to be a bit drunk to stay there. I mean like secretly. People used to bring us stuff all the time remember.[23]

Although the Picket tended to attract less overt hostility in the middle of the night, on those occasions when it was threatened by groups of individuals who were (violently) opposed to its existence, the protest could feel more vulnerable. Techniques that were successfully employed in the daytime to keep the Picket secure were not always effective on a sparsely populated overnight shift:

> Unlike daytime, where, if you get a group of troublemakers or racists, the policy of 'let's all start singing' didn't really work if there's only three of you. Quite often you'd be outnumbered. So, it was just a case of keeping quiet and putting up with it and hope that the cops on duty, if it got physical, actually did their job.[24]

To pass time overnight, picketers would often indulge in lengthy, in-depth political discussions with each other, they would sing and dance to keep warm, but they would also enjoy other pursuits that helped time pass, like games of chess or charades. In contrast to the noisy, animated theatricality of the protest during daytime and the early evening, night shifts were slower and quieter, but not necessarily passive. At these times, picketers still had to be alert to respond quickly to potential encounters as they unfolded (Bissell 2007). The fast-paced, continuous commitment to be 'non-stop against apartheid' was shaped just as much by these moments of deliberate, reflective waiting (Jeffrey 2008; Conlon 2011).

For many picketers, one of the attractions of nights spent on the Non-Stop Picket was the very eclectic and eccentric mix of people they might anticipate meeting there:

> So that was nice and you meet interesting people and crazy people and you have a laugh and then by about 4 o'clock in the morning you might just have one or two people that are so off their head they don't know how to get home, who are just sort of hanging there and gone very quiet and feeling very cold.[25]

If the odd assortment of the drunk, the homeless, the insomniac, the mad, and the disaffected could at times entertain picketers on their night shifts, they may not have appreciated at the time how important the constant presence of the Non-Stop Picket was for some of the people who bounced along to visit them. In a radio interview about how he successfully stopped using psychiatric drugs as a young man, the clinical psychologist Dr Rufus May remembered that he would:

> Walk down at night time in the rain with a bin liner to keep me dry down to the Non-Stop Picket outside the South African Embassy and talk at length with the protesters there with zest and passion and then walk home in the morning and nothing happened. You know, I could pass through these highs [there] (Rufus May, BBC Radio 4, 6 February 2001).[26]

However the night passed on the Picket, whoever was there to share banter and debate with, there was one constant that shaped the protest's night shifts – as soon as the McDonald's fast-food restaurant at Charing Cross opened at 6 a.m., a picketer would be dispatched there for coffee. Each of their comrades would then follow them, in turn, to sit for a while in the warm and try to thaw out sufficiently to stay on the picket until the morning shift (hopefully) arrived at 7 a.m. to relieve them.

Weekly rhythms

Attendance on the Non-Stop Picket peaked at the weekends. In many ways, this is obvious – more people could make time to join the protest on days when they did not have classes or work to attend. However, City Group also organised a weekly schedule of events on and around the Picket that clustered around (but were not limited to) the weekend and which could attract more participants than an average shift on the rota.

Friday evenings were the culmination of the week. Every Friday at 6 p.m. (as they had done for several years before the Non-Stop Picket started), City Group held a themed rally outside the Embassy for an hour. These rallies would respond to events in South Africa or highlight particular campaigns that City Group was supporting. They were an opportunity for the Group's supporters to come together, meet each other, and feel part of something bigger than just the (potentially) small number of fellow picketers they met during their own slot on the rota. These rallies also provided a focus for new supporters in the process of getting drawn into the Group's work. Sometimes the Friday rally only attracted a couple of dozen people; but at other times, there could be closer to a hundred people present.

If the previous week had been particularly brutal in South Africa, the Group would often place flowers on the imposing, monumental gates of South Africa House. This ritual of remembrance did not occur each week but was used, as appropriate, to remember those killed by the apartheid regime. When apartheid prisoners were executed, anti-apartheid activists were assassinated by South African agents, or the South African Defence Force massacred youth in the townships, City Group would respond to witness that act of state violence. Sometimes this would be a modest and spontaneous act by the protesters gathered on the picket for a particular shift, but during the weekly Friday night rallies or other large gatherings, a dignified queue would form as scores of protesters waited to add their flowers to the embassy gates until they were bedecked in flowers of remembrance. More often than not, the picketers would sustain the haunting melody of *Senzenina*, a song sung at political funerals in South Africa, throughout this act of witness, mourning, and remembrance.[27]

After an hour of speeches, chanting, and song, those who were not staying on the picket were encouraged to head off across Central London to the venue for City Group's weekly meeting. Inevitably, a few would slip away to the pub instead. In the early months of the Picket, the meetings were held at County Hall on the South Bank, the erstwhile home of the recently abolished Greater London Council. In later years, the meetings were held in a variety of community halls in the Kings Cross area. From 7:30 p.m. until often well after 10:00 p.m., picketers sat, debated current events in South Africa, and planned their campaigns and protests. The weekly meeting was also important in sustaining the Non-Stop Picket because it was often the best location for plugging any gaps in the Picket's rota for the forthcoming week.

Frequently, these meetings also attracted guest speakers from the South African liberation movements. Leading members of the Pan-Africanist Congress (PAC), like Zolile Keke,[28] Gora Ebrahim,[29] and Johnson Mlambo[30] (as well as members of the Azanian People's Organisation (AZAPO) and the Black Consciousness Movement of Azania, such as Strini Moodley[31] and Haroon Patel[32]), attended with some regularity (Kondlo 2009). If their presence was anticipated, or there was a major protest being planned, the numbers attending the meeting could grow significantly. In the early months of the Non-Stop Picket, these meetings regularly attracted 100 or more participants each week. That number dwindled over time.

Francis gives a good sense of the business covered in those Friday meetings, balancing what they were useful for with a sense of remembered frustration at their shortcomings:

> When I first started going they were usually quite big, with impressive turnouts, but they gradually got smaller. They were democratic but formal. Carol [Brickley] usually convened with great skill. There was a review of recent events, our financial and legal situation. Recent correspondences and news events were outlined and discussed. Importantly, the meetings were the basis from which campaigns and actions were planned. When there was a shortage of people for a particular picket shift, someone had to volunteer. People were often requested to volunteer to help with running the Group or to participate in actions. Everyone had the right to speak (though heckling and interruptions happened sometimes) ... But often the debates were very interesting, amusing and informative[33].

Like many former picketers, Francis celebrates the inclusive democracy of those meetings, in which everyone had an opportunity to speak, whilst remembering them as dragging on for too long (Polletta 2012). David Yaffe made a similar observation:

> I also used to attend quite a few of the picket meetings that used to be on a Friday night with a very, very high attendance, very, very democratic meetings, so democratic that at times you just wondered where it was all going because everybody was allowed to have their opinion and some people's opinion went on for a great deal of time. Frankly I think a lot more time was spent in those meetings than was necessary to complete the work of City Group in making decisions in a democratic forum. But that's what took place and it was unique in that respect. I don't think I've ever attended a meeting so democratic and so open as those first meetings of City Group, which went on for a long period of time.[34]

The intense shared experience of maintaining the Non-Stop Picket often provoked personality clashes, conflicts, and intense political debates that played out in those meetings. But the intensity of that shared commitment was also what led so many people to attend those meetings Friday evening after Friday evening for so long. A 'good' meeting could have a positive impact on the mood of the Picket, often throughout the following week(end), as Helen Marsden observed:

> So say for example when we did the Friday overnights, after the meetings, people would come back from meetings obviously quite buoyant and quite geed up and everything because we'd usually had lively discussions.[35]

The Friday rallies and meetings were never the only regular activities in the Non-Stop Picket's weekly schedule of events. For a while, on Tuesday evenings, the Group attempted to revive the weekly 'Picket University' that had been initiated during the 86-day picket for David Kitson in 1982 (Kitson 1987). On these occasions, either a picketer led a discussion on a particular theme, or the Group invited down a guest speaker (often a politician or journalist) to do the same. An informal political seminar would then ensue, huddled on the pavement outside the Embassy. From 1987 onwards, members of the Humanist Party collectively organised a shift led by their members (also) on a Tuesday evening.

One of the most enduring, popular, and well-supported events on the picket was the weekly women's picket on a Thursday early evening. These began on Thursday 27 November 1986, when the Picket hosted a "Viva the Women!" event to celebrate the role of women in the struggle against apartheid and as a focus for mobilising British women's groups to anti-apartheid solidarity work. This event served as the launch for a weekly focus on women in struggle that persisted on Thursday nights for the rest of the Non-Stop Picket.

The launch event was attended by about a hundred people – men and women. It was a noisy, musical event – with poetry from Jackie Kay (2010) and music from the Horns of Jericho (a lively band of street musicians). Amanda Collins from City Group spoke about her experience of sexual assault and harassment at the hands of Cannon Row Police, as well as the political campaign that City Group had waged to challenge this (see Chapter 6).[36]

From this initial launch event, the Horns of Jericho committed to play regularly on the picket every Thursday evening. Alongside these weekly pickets, the first Thursday of every month became designated for larger women's rallies. Enlivened by the Horns of Jericho's ska music, the early evening Thursday shift was often popular on the picket:

> Horns of Jericho were just great, so anarchic and a nice break from the EARNESTNESS of the Picket, as was Humanist night on the picket ... Horns of Jericho were great. I admired their commitment. But over the years they came with less people, and sometimes this main woman was on her own in the end. But this was great. "Till the walls come tumbling down" was their commitment, and they achieved that.[37]

Their shift could attract a large crowd who would dance and sing along to the music. In the process, though, the focus on women's issues could often get lost. At times, the picketers could get too absorbed in the music and having fun with their friends. This was important to building a community of protesters, but (occasionally) less effective for fostering dialogue with the passing public. At their best, these were effective times on the picket when large numbers of picketers and a lively atmosphere encouraged passers-by to stop and engage with the political message of the protest, talking to enthusiastic and buoyant protesters.

The events that structured the weekly rhythms of the Non-Stop Picket could also be quite low-key and risk going unnoticed by outside observers. To picketers, though, they could be very important. For example, at midday each Saturday, Lena Prior (a long-term anti-apartheid activist) would pull up briefly behind the Picket's banner in her Volvo estate car and deliver a box of homemade vegetarian pizzas. This was vital sustenance for picketers on a low income and attracted many to that particular shift. The Non-Stop Picket remained busy throughout the weekends (except, perhaps, Sunday mornings). The busier events provided picketers with an opportunity to get to know a broader range of people than those on their specific shifts (and, the more people they got to know, the more inclined they were to pop along spontaneously and socialise with their fellow picketers on lively shifts). These 'feature' events, which structured the weekly rhythms of the Non-Stop Picket, helped draw new supporters into an on-going commitment to the Picket and helped sustain the interest and commitment of existing members.

Sustaining non-stop activity

Sustaining a 'non-stop' protest required a non-stop commitment from a core group of activists. Although, at its peak, a hundred picketers might attend the Group's weekly meeting (out of a membership that grew to just over 1,000 towards the Picket's conclusion), in practice, a core group of only about 25–30 activists drove the Group's activities. Through their regular shifts on the picket, and the other regular weekly events and activities that they participated in, some members of this core group were devoting exceedingly long hours to their anti-apartheid campaigning. This was unsustainable – certainly without constantly recruiting and developing new activists who could share this burden of activity (or replace those who were too exhausted to continue) (Brown and Pickerill 2009). The weekly rituals, such as the Friday rallies and meetings, were useful as a mechanism to give new contacts an immediate activity to return for. However, maintaining this weekly schedule itself required an additional time commitment from activists. Inevitably, at times, these weekly events on their own were not enough to maintain the collective energy of the Group. In addition to its daily and weekly routines, the Picket developed an annual calendar of larger events, as well as means of celebrating its own longevity. Many of these annual events commemorated important moments from the history of the anti-apartheid struggle in South Africa. These larger celebrations and rallies provided an opportunity to acknowledge the commitment of regular picketers and activate new supporters to join them in sharing the responsibility of maintaining the Picket's heightened level of activity. The rest of this chapter considers the levels of commitment that activists invested in the Non-Stop Picket, as well as some of the ways this was celebrated, including through an annual calendar of special events.

A small proportion of the Group's core members committed in excess of 50 hours a week to picketing, meetings, and office work related to the Picket. While few people could sustain this number of hours perpetually, many did devote time to the organisation and the Non-Stop Picket several days (or evenings) a week. Those picketers who were members of supporting organisations, particularly the Revolutionary Communist Group, would also have commitments to branch meetings and other political duties in addition to the efforts they put into the Non-Stop Picket. Picketers from all walks of life found intricate ways to integrate their commitment to anti-apartheid protest around other responsibilities. To be 'non-stop against apartheid' was not just a commitment to long-term protest, for many picketers it was also a commitment to a particular pace and intensity of campaigning.

The determination and commitment to maintain a continuous protest (until Mandela was released) was not just expressed in the amount of time that picketers spent engaged in campaigning. Their determination to keep going was also witnessed in the tenacity with which they ensured nothing disrupted the continuity of the Non-Stop Picket. On the night of 15/16 October 1987, England was hit by hurricane force winds. It was the worst storm for nearly 300 years. Across the country, millions of trees were felled by the winds, there was significant damage to buildings, and several people were killed by falling debris. Despite this, City Group activists maintained their pledge to keep a non-stop picket going outside the South African Embassy.[38]

It is important not to romanticize the resolve of the non-stop picketers too much, though. At times, particularly in the last year of the Picket, morale and discipline were low, people would fail to turn up for their shifts, and the continuity of the Picket was maintained by a thread. At the same time, the response of a small number of picketers to the extraordinary events of that October night in 1987 demonstrate how seriously City Group members took the integrity of the Picket and their pledge to maintain it until Mandela's release.

Annual cycles of protest

From the beginning of the Non-Stop Picket, City Group recognised the need for regular larger events to sustain the interest and morale of its members. Some of these larger demonstrations were specific to campaigns that the Group was running at the time, such as the campaign to Free Moses Mayekiso (Bell 2009), or the Upington 26 (Durbach 1999). Some of these dates were already fixtures in City Group's annual calendar of activities; but, as the Non-Stop Picket went on, new cycles of annual protests were initiated and repeated.

Over the four years of the Non-Stop Picket, certain events from recent South African (and Namibian) history were regularly marked there. On 26 June each year, they marked (what was then known as) South African Freedom Day, remembering the adoption of the Freedom Charter by the

African National Congress (ANC) and their allies on that day in 1955 (Lester 1998). Each August, the Group celebrated South African Women's Day, and most Decembers, they also celebrated Namibian Women's Day. In September, they remembered the death in police custody of Steve Biko from the Black Consciousness Movement (Biko 2015; Gibson 2011). But two of the largest events in the calendar were regularly the 16 June anniversary of the start of the school students' uprising in Soweto in 1976 (see Chapter 7) and the commemoration of the 1960 Sharpeville massacre.

On 21 March 1960, 69 unarmed protestors were shot dead by South African Police at Sharpeville (Lodge 2011). They were participating in a demonstration called by the PAC against the pass laws, a central facet of the apartheid regime, which regulated where non-whites could live and work. The Sharpeville massacre was a turning point in South African history and led to a chain of events that shaped the direction of resistance to apartheid both in South African and internationally (Gurney 2000).

For City Group, like many anti-apartheid campaigns around the world, 21 March was a key date in the protest calendar. The date was particularly important for City Group, given their 'non-sectarian' stance of supporting all anti-apartheid liberation movements, as it provided an opportunity to consolidate their relationship with the London representatives of the PAC. Unusually, given the centrality of the Non-Stop Picket of the South African Embassy to City Group's work at the time, in the late 1980s, Sharpeville commemoration events normally took place away from the Embassy itself. In 1988 and 1990, indoor rallies were held at Conway Hall in Holborn; while in 1989, the "Remember Sharpeville, Remember Langa" rally was held at the London School of Economics. These evening meetings were invariable addressed by senior leaders of the PAC. They could attract several hundred people and normally culminated in a candle-lit march from the venue to the Non-Stop Picket.

As important as these events were, the Non-Stop Picket's annual calendar of events was not just structured by South African history. Over time, the Picket developed a cycle of events that responded to other (political) events around it and proactively sought to build the Picket on days when, for one reason or another, its continuity might be strained or threatened. In the rest of this section, we chart two very different events that became part of the annual calendar of the protest.

Remembrance Sunday

Remembrance Sunday was always a tense day on the picket, which was less than a mile away from the Cenotaph on Whitehall, the focus for state rituals of remembrance. Sometimes the platoons of former servicemen and women passing the Picket on their way to or from the ceremony at the Cenotaph could be antagonistic towards the anti-apartheid cause. Those retired soldiers were never the real problem, though. The threat posed to the Non-Stop

Picket on Remembrance Sunday was the annual parade that passed the Cenotaph later in the day by the motley membership of the neo-fascist National Front and/or British National Party. They definitely did not approve of anti-racists protesting against apartheid and had a tendency to either try to launch full-on assaults against the Picket or, more 'subtly', attempt to provoke violent arguments with picketers.

Some picketers remember the tension and fear that could grip the Picket on Remembrance Sunday:

> On Remembrance Day there was a bit where you were actually shitting yourself because you thought you were going to get killed that day, where you thought no actually, we have got enough people to defend ourselves and if we are strong in number we can defend ourselves.[39]

In response to the potential danger of neo-fascist attack, each year, on Remembrance Sunday, large numbers of City Group supporters and other anti-fascists would mobilise, gathering on the picket to defend it against potential attack. On 13 November 1988, in particular, hundreds gathered outside the Embassy in response to the specific threat of fascist attack.

In parallel to these mobilisations, activists from the Non-Stop Picket would often patrol the area with Anti-Fascist Action (Hayes 2014) to identify and head off groups of fascists heading towards the Picket. These anti-fascist mobilisations to defend the Non-Stop Picket once again highlight the ways in which, for City Group members, the politics of anti-apartheid solidarity activism could never be separated from anti-racist and anti-fascist work in Britain. Although this annual mobilisation was important to the continuity of the Non-Stop Picket, it was also an opportunity for City Group to maintain a connection with sections of the British Left that did not always prioritise anti-apartheid campaigning.

Christmas on the Picket

The Non-Stop Picket was 'non-stop', and that included Christmas Day. Although it was never their primary motivation, by resolutely continuing their protest over the festive period, the Non-Stop Picket often attracted a certain level of media coverage.[40] This was more than a measure of the picketers' determination; for many picketers, especially those who lived far away from their families, or had strained family dynamics, spending Christmas Day on the Non-Stop Picket offered them the opportunity to be with the people who really mattered to them. But some families also made it part of their Christmas rituals for those few years, as Kathy recalled:

> [On] Christmas Day my daughter and I would come down and do a shift on the Picket. My daughter was about what 12 or 13 then or maybe a bit older.[41]

One tradition that developed on the picket was the annual visitation on Christmas Day by 'Father Freedom'. In 1986, the Picket's first Christmas, this role was taken by the Mayor of the London Borough of Islington, Bob Crossman. In a letter dated 8 December 1986, Andy Higginbottom wrote to Bob Crossman inviting him to attend the Picket on Christmas Day:

> Our idea is that yourself and Mary Cane [Mayor of Camden] act as a socialist Father and Mother Christmas, preferably Father and Mother Freedom and we have a present giving on the picket to the youngsters who spend a lot of time there keeping it going.[42]

Jeremy Corbyn, then (and still) the Member of Parliament (MP) for Islington North, remembers attending the Picket on Christmas Day with Bob Crossman and others:

> I made a point of visiting them on Christmas Day with the Mayor of Islington, with the late great Councillor Bob Crossman. So we spent – Bob Crossman (who was the Mayor) and his partner, Martin McCloghry – we spent the day visiting the hospital, [...] the prison, two pensioners' lunches. We didn't get anything to eat by the way, we just went to say 'hello' to the pensioners having their lunch, and then went down to Trafalgar Square for the Non-Stop Picket with the Mayor's chauffeur driving us around.[43]

In subsequent years, the role of Father Freedom was played by Hamilton Keke, the Chief Representative of the PAC in London (Kondlo 2009), and David Kitson. The two former South African political prisoners played the role wearing a traditional Santa costume that was edged with black, green, and gold braid and adorned with badges and slogans from the anti-apartheid struggle.

Although Christmas time on the picket could be fun and brought in extra financial donations to sustain the Picket and material aid to send to South Africa, in other ways, it was business as usual. On the evening of Christmas Day 1986, Grace Livingstone was Chief Steward and recorded the following note in the Steward's book:

> It's been quite a peaceful evening until about half an hour ago AD464 [a police officer] came on, insulted a few people on the picket a few times, then walked off. Martin was leaving the Picket, AD464 met him on the corner, searched him and then threatened to beat him up when he was in plain clothes. He called Martin 'a disease' and insulted him more and other people on the picket. He said he was going to cure the ['disease'] of the Picket.[44]

Even during the festive season, the war of attrition between the Non-Stop Picket and the Metropolitan Police continued. For the police from Cannon Row Police Station, there was little Christmas cheer for anti-apartheid protesters.

In addition to celebrating key dates from the history of the anti-apartheid struggle, the Non-Stop Picket developed an annual cycle of activities that helped focus their commitment to remain until Mandela was released from gaol. They quickly appreciated the importance of making a feature and a (positive) fuss of those days that threatened that commitment or that, like Christmas Day, could have been difficult to sustain otherwise.

Celebrating longevity

In addition to its annual cycle of protests and commemorations, the Non-Stop Picket also organised special events to mark and celebrate its own longevity and achievements. In September 1986, City Group marked the first 100 days of the Non-Stop Picket by issuing a report to the media on '100 days of police harassment', which provided members with an additional cause to campaign around.[45] More usually, significant threshold dates, such as the 500th and the 1000th days of the Non-Stop Picket, provided an excuse for a celebration and a larger than usual crowd outside the Embassy. Similarly, each April, the Group celebrated the anniversary of the start of the Non-Stop Picket with a sizeable and frequently militant birthday rally. In this section, we examine in more detail the role played by those anniversary protests and the Picket's 1000th day in January 1989.

Anniversary celebrations

The celebration of the first year of the Non-Stop Picket (unlike those in subsequent years) was held on the anniversary itself. On 19 April 1987, 400 people came together outside the South African Embassy to celebrate a year of continuous anti-apartheid protest there. During that year, the Picket had collected over 300,000 signatures on the petition calling for the release of Nelson Mandela. In that time, City Group had donated £5000 to political prisoners and their families. But, City Group spent at least double that in legal fees defending the more than 100 people who had been arrested at their protests (see Chapter 6).

The first year of the Picket had been hard-won, but its celebrations did not go unchallenged. During the rally, a squad of Territorial Support Group officers from the Metropolitan Police attacked the Picket while trying to confiscate a small, 'unauthorised' stage from which the speakers were addressing the crowd. In the process, Steve Kitson was knocked unconscious. Steve was dragged off to a waiting police van, still unconscious. In uproar, picketers demanded that an ambulance be called to attend to him. After considerable delay, he was taken to hospital, where he was kept under medical observation overnight. He was not charged at the time, but for good measure, the police insisted that he report to Cannon

Row Police Station a week later to find out if charges would be pressed against him.[46]

During the scuffles that led to Steve's injury, one officer lost his helmet. Angered by the injuries sustained by her son, Norma Kitson later displayed the police helmet on a pole from which an ANC flag also flew. Perhaps understandably, the police would not tolerate this for long, and a further ten picketers were arrested – one, slightly ludicrously, for assaulting a police officer with a daffodil. Despite the tensions and barely controlled anger of that day, picketers later immortalized the events of that afternoon in a humorous song – the novelty of helmet on a pole and an arrest for assault with a daffodil were always likely to enter Picket folklore (Figure 4.1)[47].

The Picket's second anniversary in April 1988 was marked with a march to the Embassy from Marble Arch near Hyde Park – a traditional route for political demonstrations in London. Something of the flavour of that demonstration is captured in a report on the event that Jon Kempster

Figure 4.1 Arrest of protester during first anniversary rally, 19 April 1987, photographer unknown. Source: City Group.

recorded for the *Black Londoners* programme on BBC Radio London. Jon began his report by offering some contextual information to his listeners:

> About one thousand people came from all over Britain to mark the second anniversary of the Non-Stop Picket, which has remained outside the South African Embassy every day and night since April 1986.[48]

There followed two excerpts from speeches made by members of the South African liberation movements. David Kitson was heard to explain:

> Comrades and friends we are here today to celebrate two years of the Non-Stop Picket outside South Africa House. And believe you me, every freedom fighter in South Africa and every political prisoner in South Africa is watching this Picket keenly because they regard it as a part of their struggle for liberation. There was once a non-stop picket before in 1982 run by the City of London Anti-Apartheid Group. That was when I was in gaol, and after 86 days we were moved to a better prison and the prison authority told me that we had been moved because of requests from the South African Embassy who were irritated by the picket outside its gates[49].

His words were greeted with loud cheers from the crowd. David Kitson's sentiments were echoed by Molefe Pheto, who explained why he was there to mark the second anniversary of the Picket:

> I am here on behalf of the Black Consciousness Movement of Azania, which is one of the liberation movements from that part of the world. I stand here to congratulate politically the City Anti-Apartheid Movement for their courageous stand in the campaign for the Sharpeville Six. But the struggle is not over yet. This is not the end, it is only the beginning and we want to warn you, we who come from that part of the world, that this is serious business. You have only started, so gird your loins for much more heavy work, but the struggle is escalating in that part of the world.[50]

At a certain point during the rally outside the Embassy that afternoon, the protesters' attention to the scheduled speakers was interrupted – a cheer roared through the crowd, and their attention shifted elsewhere – in quite the opposite direction, in fact. Two City Group activists, Gary Rose and Dave Kenny, had climbed scaffolding in front of the National Gallery on the north side of Trafalgar Square and unfurled a banner calling for the release of the Sharpeville Six (anti-apartheid activists who were on death row in South Africa at the time).

Direct actions like this, during large rallies, played a similarly affective role to the political speeches. They were expressions of anger at apartheid, but they were intended to inspire protesters to take action themselves. It worked on this occasion – when a second duo were arrested on the roof of the gallery, the crowd responded by engaging in a large sit-down protest on the roadway in front of the Embassy. This didn't happen with great regularity, but neither was it an unheard of occurrence. Once or twice, the police took a tactical (or perhaps a political) decision to allow these sit-downs to run their course, rather than make mass arrests. That was not the case in April 1988. The police responded quickly and efficiently in clearing the road, arresting 31 protesters for highway obstruction in the process. For City Group, like other social movements, such acts of civil disobedience were both a show of strength, an attempt to make a political point by disrupting normal business on the city's streets, and an opportunity to potentially gain media coverage for their cause. Indeed, the *Black Londoners* radio report ends with Jon Kempster explaining that the picketers ended their rally with a sit-down protest in the road outside the Embassy.[51] By blocking the road, they disrupted the flow of people and traffic around Central London, using the inconvenience of this disruption to the city's infrastructure and logistics for modest political leverage and hoping that resulting media coverage might amplify their message (Graham 2010).

The major rallies, like those that took place on or around the anniversary of the start of the Picket, were an opportunity for the supporters of City Group to come together and collectively renew their commitment to anti-apartheid protest – often through acts of direct action or support for those taking such action. These acts held the picketers together and kept them focussed on their cause (Figure 4.2).

Amongst an impressive battery of speakers, the third anniversary rally in April 1989 was read a message of support from Zephania Mothopeng, President of the PAC. He had been released from jail in South Africa late the previous year due to his deteriorating health and had arrived in London a few days before the rally to undergo medical treatment, but was too ill to attend the rally himself. His message to the rally stated:

> Three years have passed since you launched your Non-Stop Picket in which you focus on the important theme – the unconditional release of all political prisoners in Azania.
> You have in the course of this 1,100 days and nights also rallied for the closure of the racist embassy. Comrades, your militant revolutionary task has made you targets for vicious forms of harassment and attack, but you have been undeterred. Moreover your determination and sacrifice have been crowned with important gains, which include my own unconditional release as well as the reprieve of the Sharpeville Six.[52]

Unsurprisingly, this positive endorsement from a leading figure in the anti-apartheid struggle provided a powerful morale boost to City Group

94 *Being non-stop against apartheid*

Figure 4.2 Rally for the third anniversary of the Non-Stop Picket, April 1989. Source: Gavin Brown.

activists to continue their on-going picket of South Africa House into its fourth year.

1000 days and nights

On Saturday 14 January 1989, City Group celebrated the 1000th day of the Non-Stop Picket with an ambitious 12-hour rally. The actual 1000th day of the Picket had been two days earlier on 12 January, and on that day, a delegation from the Group delivered half a million signatures on their petition to Downing Street calling for Mandela's unconditional release.

The picketers had been building for this celebration for months. Stickers and posters advertising the rally had been posted around London, thousands of leaflets had been distributed, and the Picket itself was adorned with a special banner in the weeks running up to the event, which had changeable numbers on it, to allow a countdown to the 1000th day. In the office, activists had been busy contacting musicians and celebrities who they thought might be persuaded to lend their support to the rally (Davies 2012).

> We used to scan the newspapers, anyone who said anything anti-racist we'd write to them ask them if they wanted to come to the Picket, and we got a lot of good messages and people coming down as a result of that as well.[53]

The 1000th day rally served several functions. It was a celebration of the stamina and determination of the picketers. And it was an opportunity to use this story of endurance to gain publicity both for the Picket itself and to draw renewed attention to the struggle against apartheid in South Africa.

The 1000 days rally *did* attract considerable media attention, including a live broadcast from the Picket by Soviet television. Perhaps not surprisingly, many of the media reports focussed on the 'human interest' aspects of life on the picket, profiling individual picketers, as much as they addressed the political reasons for the Picket's existence. One report in the *Guardian* focussed on 75-year-old Rene Waller's commitment to the Picket as its human-interest story.[54]

If sustaining the Non-Stop Picket for a thousand days and nights was an achievement, then sustaining the celebration rally for 12 hours was also a challenge. To help structure the rally, and to provide a focus for mobilising different constituencies amongst City Group's supporters, each of the Group's themed sub-groups took responsibility for organising specific periods of the day. Dominic, as Youth & Students Organiser at the time, took the lead in organising an hour, mid-afternoon, themed around "Youth and the struggle against apartheid."[55] While some of those he invited to speak did help attract publicity, their contributions were not without some problems:

> I remember the thousandth day, partly because I organised part of it, and that was very busy and it was like very busy for hours and hours and hours and I remember one of the people I asked to speak who came was Genesis P-Orridge of Throbbing Gristle and Psychic TV… And they wanted to do a benefit actually, but it never happened. They put our address on some of their records at the time said give, send money. I think they put Save the Rhino and City AA, and they were sort of countercultural folk, but his speech was nuts and it was sort of all obsessed with Terre'Blanche and the AWB,[56] it was kind of embarrassing. But he was sort of quite well known and that helped to get sort of things in the music press that it was going on.[57]

There were political speeches and messages of solidarity from representatives of the liberation movements in South Africa, from British politicians, and from campaigning organisations that supported the Picket.

The Picket was all about engaging people in action against apartheid, so throughout the day, there were activities and low-level direct actions to offer participants something to do. Several times during the day, the rally went mobile and marched around the perimeter of South Africa House. A thousand black balloons, one for each day of the Picket, were released in a highly visual celebration of the Picket's duration.

City Group approached the 1000-days rally as an opportunity to attract new resources and energy to its campaigning. On the day, pledge cards were circulated through the crowd, which encouraged all those attending to sign up for a regular shift on the picket rota. The front page of the pledge card displayed a photo of Nelson Mandela accompanied by a quote from him: "I cannot and will not give any undertaking at a time when I and you, the people are not free. Your freedom and mine cannot be separated. I will return."[58] Mandela's commitment to the struggle against apartheid was used here both as a reminder of the Picket's purpose, but also to encourage sympathisers to follow his lead and make a commitment of their own to the Picket. The words "I will return" were printed in bold, encouraging protesters to return themselves.

The publicity surrounding the rally and the success of the event itself served as a morale booster for regular participants in the Picket, and the increasing number of rota pledges (and standing order donations) solicited at the rally breathed new life and energy into City Group's work. The breadth of the people attracted to the celebration rally also had a powerful emotional impact on how young activists framed their commitment:

> And having such, I think there's a diversity of peoples, that's what made it feel that it was possible, because you just think well no actually it's not just 17/18-year-old girls who feel like this, it's not just – this isn't just a phase we're going through, because there's pensioners here, there's firemen here, there's lawyers here, there's, everybody's here, so we all do feel like this, and that's what makes it feel possible.[59]

Many of the larger rallies and demonstrations that City Group organised were reactive. They responded to events in South Africa, whether those were uprisings in the townships, repressive crackdowns by the state, political trials, or executions of political prisoners. The anniversary rallies on the picket (along with the events that marked 100, 500, and 1000 days of the protest) were different. They were a celebration of what anti-apartheid protestors had achieved on the streets of London.

Being (youthfully) non-stop

Being 'non-stop against apartheid' was about more than a commitment to an on-going, continuous protest; it also implied a particular pace and intensity of political activity. Whatever their background, time spent on the Non-Stop Picket was always fitted around the rhythms of participants' lives. Hema, a single mother in her late 20s, outlined the more complicated logistics involved in making time for her anti-apartheid campaigning:

> Daughter to childminder, me to work (I was a newly qualified maths teachers), pick up daughter, come home, then either maybe a shift on the Picket or the weekly meeting (Fridays, I believe). I did some painting, flag-making etc. for the Picket and would sometimes go fly-posting with [Claus].[60]

As picketers became increasingly committed to the Non-Stop Picket – politically, but also socially and emotionally – many also increased the amount of time they gave each week to different activities and practices associated with anti-apartheid solidarity work and maintaining an on-going street protest.

> I would do at least one regular shift on the picket, I would often do 'overnights'. I think there were committee meetings on Wednesdays and then the full City Group meetings on Fridays and I would attend both of those. I would be at the office formulating press releases to cover any 'extra picket demonstrations' and would take turns in the office, phoning police stations etc. in the event of any of our members being arrested.[61]

> Yeah, my typical week would have been my Monday shifts, my Wednesday shifts, Friday meetings, Saturdays, sometimes office work in between, and as you know my partner for five years was very involved as well, so it was kind of nearly all encompassing. Well it was a lot. And, it wasn't just the stuff that was on the picket; there was the stuff that we did at the airways, the stuff that we did in the trolley pushes. It was staying in police stations for hours waiting for people to be released after actions. Yeah, so it was a lot, yeah.[62]

The level of activity undertaken by Sharon and Deirdre was not exceptional for young picketers. Throughout the Picket, a core group of activists (perhaps no more than 30 at any one time) engaged in some anti-apartheid activity more days than not. Most, but by no means all, of those devoting this time were in their teens or early twenties. For some of them, this level of commitment could be arranged around their work and other commitments:

> I think I did a couple of shifts, but even when I wasn't on the shift I'd go down pretty much every day. Because as I say I was working around the corner, so it was like a draw constantly.[63]

For others, in the end, their commitment to the Picket dislodged all other commitments. Many chose unemployment in order to facilitate their anti-apartheid activism:

> I don't remember the exact details but I more or less spent most of my time there, having signed on after giving up my child-minding job. It became like a job really.[64]

This level of commitment was not just the preserve of young picketers. There were 'non-stop' pensioners like Rene Waller (mentioned earlier) who could also make a very regular commitment to the Picket or attendance at City Group's office and meetings. Sustaining that level of daily involvement was less easily sustained for those adult picketers who had careers (rather than more casual jobs) – and yet several of them tried to make that commitment.

> I would be on the Picket most days. Not every day, but ... I had a full-time job at the time, but I would try, certainly when it started, I have to say I tailed off, I couldn't keep it up, but at the beginning I would try to do one night shift a week, normally at the weekend. Occasionally I would do it during the week.[65]

Although thousands of people stopped and spent time on the Non-Stop Picket during its existence, in practice, to maintain its continuity, a small, dedicated group of core activists made themselves 'non-stop against apartheid'. This core group involved people of all ages, from their mid-teens to their 70s, but the majority of them were either teenagers or young adults under the age of 25. These youthful activists had the physical energy to commit long hours to protesting against apartheid. The circumstances of their lives – in large part shaped by their youth – meant that few had the types of personal or professional commitments that prevented them from finding the time to put into their activism and the flexibility to adapt their schedules at short notice to respond to events as they unfurled. City Group acknowledged the commitment of students and unemployed picketers and would pay (from the donations received on the Picket) the travel expenses of unwaged picketers who had more than one regular shift on the picket each week and engaged in other activities for the Group.

Young activists (and others) may have had the time and energy to invest in their campaigning, but being 'non-stop against apartheid' could become physically and emotionally unsustainable at times. While the existence of the Non-Stop Picket relied on the enthusiasm of young picketers, it also required a constant influx of new activists in order to keep going. The Picket operated through a weekly cycle of activities that provided existing activists an opportunity to meet with each other and feel part of a larger whole, as well as an opportunity for new supporters to be inspired to do more. If someone had been impressed by the commitment of a handful of protesters standing outside the embassy gates on a wet Tuesday afternoon, how much more inspired might they be by the presence of 50 people at the Friday evening rally? This cycle of larger events was also replicated on an annual basis, as City Group organised mass protests to commemorate key events in South African history and celebrate the Picket's own growing longevity. There was seldom more than six weeks between these larger events, and at certain times of the year, they occurred even more frequently. To organise, publicise, and run these larger events took additional time and energy beyond the perpetuation of the Non-Stop Picket. But, like the events themselves, the process of organising these rallies could be fun. Young picketers frequently felt a great sense of achievement when events they had helped to organise came, successfully, to fruition. This regular calendar of events throughout the year provided an opportunity for City Group's more peripheral supporters to maintain contact with the Non-Stop Picket. Because they tended to be loud, vibrant, and visually interesting events, and frequently contained some element of confrontational direct action against the Embassy, these were intensely affective events that could inspire new contacts to commit their time to the Non-Stop Picket or revitalize the energies of existing activist.

Notes

1. In City Group's folklore, this phrase is often claimed to have been used by the (Communist Party of Great Britain affiliated) journalist Seumas Milne in an article in *The Guardian* criticising City Group and the RCG in their political conflicts with the leadership of the AAM. We have checked the four articles Milne wrote about City Group's expulsion from the AAM and the phrase does not appear to be used in any of them (although he does comment on their 're‐markable tenacity'. This is not to say that the backhanded compliment was not levelled at City Group at this time, possibly by Milne, but it does not appear to have been published in *The Guardian*. S. Milne (1984), 'Left takeover bid for Anti-Apartheid', 27 October, p. 2; S. Milne (1984), 'Anti-Apartheid's leaders defeat ultra-left move', 29 October, p. 28; S. Milne (1985), 'Apartheid group expels 'mavericks'', 25 February, p. 3; S. Milne (1985), 'Separate developments in common', 23 March, p. 17.
2. In the interests of transparency, it is worth noting that Gavin Brown held this position for approximately the last five months of the Non-Stop Picket and several months after that.
3. Interview with Amanda Collins.
4. City of London Anti-Apartheid Group (1990), *The Non-Stop Picket: A User Friendly Guide*. Unpublished internal document, p. 1.
5. Interview with Richard Roques.
6. Many of the 'stewards books' survive amongst the papers that were packed away when the City Group office was closed down in 1994. Where there are gaps in the set, we assume this is because specific notebooks were passed to defence solicitors, as evidence, following arrests.
7. Interview with Georgina Lansbury.
8. Interview with Andre Schott.
9. Interview with Ann Elliot.
10. Interview with Trevor Rayne.
11. Interview with Lorna Reid.
12. Interview with Simone Maloney.
13. There is, of course, a certain incongruity in protesters who sought to close down the South African Embassy, feeling as if, at times, they were guarding South Africa House.
14. Interview with Mark Farmaner.
15. Interview with Ann Elliot.
16. Interview with Deirdre Healy.
17. The lyrics of the song commemorating Samora Machel were:
Sleep well Samora
Sleep well Samora Machel
Sleep well Samora
Sleep well Samora Machel
He was a communist fighter

> He was a communist leader
> Sleep well Samora Machel

18 Interview with Jacky Sutton.
19 Interview with Penelope Reynolds.
20 Email correspondence with Sally O'Donnell.
21 Interview with Helen Marsden.
22 Interview with Nick Manley.
23 Interview with Gary Lowe.
24 Interview with Nick Manley.
25 Interview with Mark Farmaner.
26 R. May (2001), *Taking A Stand*, [Radio], BBC Radio 4, 6 February. BBC copyright content reproduced courtesy of the British Broadcasting Corporation. All rights reserved.
27 The song contains lyrics in Xhosa, Zulu, and, in some versions, English. Its mournful title translates as "what have we done?" The song was included in: City Group Singers (1987), 'Senzenina', from *Freedom Songs from the Release Mandela Non-Stop Picket, 1986–1987*, London: Aluta. https://soundcloud.com/gavinbrown/senzenina (Accessed 24 August 2016).
28 Zolile Hamilton Keke was a political prisoner on Robben Island for ten years, from 1963 to 1973. In December 1977, he was the youngest defendant in the Bethal Treason Trial. Having received a suspended sentence at that trial, he chose political exile and left South Africa. He was the Chief Representative of the PAC in the UK and Ireland from 1982 to 1985. He also served as Chief Representative in Libya and Iraq, and was a member of the PAC's Central Committee (Kondlo 2009). He died in February 2013.
29 Gora Ebrahim joined the PAC in 1957 and went into exile in 1963, becoming one of its most senior members of Indian heritage. He joined the Central Committee of the PAC in 1981, when he also became the organisation's Director of Publicity and Information (Kondlo 2009). In December 1981, he became the PAC's Permanent Observer representative at the United Nations (UN), working closely with the UN Special Committee against Apartheid. Following the unbanning of the PAC in February 1990, he became the PAC's Secretary for Foreign Affairs. He became increasingly disillusioned by the internal disorganisation and feuding within the PAC during the 1990s and joined the ANC shortly before his death in 1999.
30 Johnson Mlambo was a founding member of the PAC. He served 20 years on Robben Island and left South Africa on his release in 1983. In exile, he first served as Secretary for Foreign Affairs and then, from 1985, as Chairman of the PAC. He effectively led the PAC in exile from 1985 to 1990 (Kondlo 2009). He served as Deputy President of the PAC from 1990 to 1994.
31 Strinivasa ('Strini') Moodley was a close associate of Steve Biko and was involved in various Black Consciousness organisations from the early 1970s onwards. He was briefly a regional organiser for the Trade Union Council of South Africa in 1970 and became the Publicity Director for the South African Student Organisation (SASO) in 1973. During the Durban strikes of 1973, he was banned under the terms of the Suppression of Communism Act. After an arrest in 1974, he was convicted under the Prevention of Terrorism Act and sentenced to six years imprisonment on Robben Island. Following his release in 1981, he joined AZAPO and worked as a journalist, serving on the national executive of the Media Workers Association. He died in 2006 on the 12th anniversary of the first post-apartheid elections in South Africa.
32 Haroon Patel was a leading member of AZAPO and the Workers Organisation for Socialist Action (WOSA), a Trotskyist organisation, before joining the Muslim Youth Movement in the early 1990s (Jeenah 2015).

33 Interview with Francis Squire.
34 Interview with David Yaffe.
35 Interview with Helen Marsden.
36 A speech was also given by Didi Brizi from Wages for Housework. They, along with other radical anti-racist feminist groups such as Southall Black Sisters and Camden Black Sisters, were long-time supporters of the Non-Stop Picket.
37 Interview with Claus.
38 M. Burgess (1987), 'Hurricane? Business as usual!', *Non-Stop Against Apartheid*, No. 24 (December), p. 7.
39 Interview with Helen Marsden.
40 The socialist daily newspaper, *Newsline*, published a photo of three picketers under the headline "The Free Mandela Christmas picket" in December 1988 (undated and un-paginated press cutting in City Group papers).
41 Interview with Kathy Fernand.
42 Letter from Andy Higginbottom to Councillor Bob Crossman, Mayor of London Borough of Islington, 8 December 1986.
43 Interview with Jeremy Corbyn MP.
44 Entry in Non-Stop Picket Stewards' notebook by Grace, 25 December 1986.
45 City of London Anti-Apartheid Group (1986), *100 Days and Nights: A Record of Police Harassment* (28 July). See also: C. Brickley (1986), '100 days and nights – racist police attack picket', *Fight Racism! Fight Imperialism!*, 62 (September), p.5. See also Chapter 6.
46 The author Lynne Reid Banks, who was a personal friend of Norma Kitson, reported these events in an article that she wrote for *The Observer*. There is also fictionalised account in her novel, *Fair Exchange* (Reid Banks 1998). See: L Reid Banks (1986), 'The day I lost my illusions', *The Observer*, Weekend section (21 June), p. 49.
47 The picketers generated their own lyrics to the tune of *Spitting Image's* 'Chicken Song' (itself a parody of Black Lace's summer holiday hit "Agadoo"):
 Hold a tyre in the air
 And a match in the other hand
 Get the petrol out and then burn the racist rand
 Support the striking miners and blow up a train
 if you do it wrong
 You can start it all again...
 The lyrics evolved over time, charting some of the more ludicrous arrests on the Picket, including *"throw a daffodil and get charged with police assault...."*
48 J. Kempster (1988), 'Report on the Non-Stop Picket of South Africa House' (excerpt from *Black Londoners*, 18 April), *BBC Radio London*. A recording of this report is deposited in the British Library National Sound Archive, C1499/1 C2. BBC copyright content reproduced courtesy of the British Broadcasting Corporation. All rights reserved.
49 J. Kempster (1988), 'Report on the Non-Stop Picket of South Africa House'.
50 Ibid.
51 Ibid.
52 ZL. Mothopeng (1989), 'Message of solidarity from the President of the Pan Africanist Congress of Azania', *Non-Stop Against Apartheid*, No. 34 (June), p.5.
53 Interview with Helen Marsden.
54 D. Sharrock (1989), 'Apartheid pickets celebrate 1000 days', *The Guardian* (13 January).
55 Alongside the youth and students section, different periods during the rally focussed on the experiences of trade unionists, women, and lesbian and gay activists

in fighting apartheid, all organised by City Group's different sub-groups. Late in the afternoon, the focus was "Smash Apartheid, Smash Racism".
56 The Afrikaner Weerstands Bewewing (Afrikaner Resistance Movement) or AWB, led by Eugène Terre'Blanche, was an extreme Right-wing paramilitary movement in South Africa that opposed any reform of apartheid and campaigned for Afrikaner separatism (Guelke 2005).
57 Interview with Dominic Thackray.
58 These words come from a statement read out by Mandela's daughter Zindzi at a UDF rally held to celebrate Archbishop Tutu's award of the Nobel Peace Prize, at the Jabulani Stadium in Soweto, on 10 February 1985.
59 Interview with Helen Marsden.
60 Interview with Hema Patel.
61 Interview with Sharon Chisholm.
62 Interview with Deirdre Healy. See Chapter 5 for a discussion of direct actions away from the Non-Stop Picket.
63 Interview with Georgina Lansbury.
64 Interview with Simone Maloney.
65 Interview with Andy Higginbottom.

5 Defending the right to protest

The Non-Stop Picket actively sought to disrupt the business of the South African Embassy. This chapter examines the Picket's relationship with the police and examines how picketers took action to defend the existence of the Non-Stop Picket and their right to protest against apartheid in the ways they chose.

The presence of the Picket was contested by the South African Embassy and the Metropolitan Police; and picketers struggled to maintain their presence there, often establishing new precedents around civil liberties and the right to protest that had lasting national consequences (Bailey and Taylor 2009). As Carol Brickley explained to Radio London's listeners on the occasion of the Picket's second anniversary:

> The main obstacles to us being here actually have been the British police and their interpretation of the law and the right to demonstrate and on a number of occasions they've attempted to move us from outside the Embassy and prevent us from demonstrating here despite the fact that we do have the right to be here under British law. A sort of attempted wearing down process, if you like.[1]

Key points of contention between the Picket, the police, and the Embassy are examined in this chapter (drawing on our interviews with retired police officers, as well as picketers). In particular, this chapter examines how, through a two-month campaign of civil disobedience, picketers regained the right to protest directly outside the embassy gates, after the Metropolitan Police forcibly moved them in May 1987, repeating the success of the South African Embassy Picket Campaign in similar circumstances in 1984 (see Chapter 2).

Points of contention

Positioned on the pavement directly outside South Africa House, the Picket was strategically placed (Bosco 2006) to draw attention to apartheid and bring pressure to bear on the regime's representatives and allies in the UK.

The Picket did not just stand outside the Embassy bearing witness to apartheid's crimes; it took direct action against apartheid's representatives. This inevitably brought picketers into conflict with the police, courts, and other representatives of the British state. These experiences of arrest, harassment, surveillance, and, at times, police violence led many young activists to question the role of the state.

Throughout the 46 months of the Non-Stop Picket, the Metropolitan Police tried many ways of stopping City Group from protesting effectively outside the South African Embassy. In doing so, they were responding to political pressure applied to the Foreign and Commonwealth Office by the South African Embassy itself.

In the early months of the protest (in 1986), the Picket fought hard to defend where it could stand, how much space it could occupy, and how picketers could engage with the public. Carol Brickley remembers one particular incident that was indicative of the way in which the Group responded with humour and contempt to these attempts to limit their protest:

> There were constant struggles over all sorts of things. I remember at one point the police decided that they would allow us to put down on the pavement only a box the size of A4. If we put anything else down they would take it away. So what we did is we arranged for a lot of rubbish on the picket, I mean large quantities of rubble effectively, which we then refused to move, which they then had to take to Cannon Row police station and store as our property. They used to send me letters about this rubbish, saying they were storing it and what did we want to do with it. We wrote back saying we hoped that they were keeping it carefully.[2]

One of the key points of contention was how vocally the picketers could make their presence felt. David Gilbertson, a Chief Inspector responsible for policing the Picket at the time, recalled:

> There was continual noise, from trumpets, tambourines, and small drums, (euphemistically referred to as 'music') from the demonstrators, which was greatly disliked by the South African diplomats in the building. They also used loudhailers to address passers-by, which became a real issue with the South African officials.[3]

The Picket *was* noisy – megaphones were used to engage passing members of the public in debate about apartheid; but, in the process, their amplified sound was also used to disrupt the 'peace and dignity' of the South African Embassy. Within ten days of the Non-Stop Picket starting in April 1986, the Embassy had complained to the Foreign and Commonwealth Office about the noise:

> – (South African Embassy official) said that the demonstration was an impairment of the dignity of the Mission; the noise from the loud hailer

prevented the Embassy staff from properly conducting their business. – (FCO official) stressed the Secretary of State's personal interest in the matter and our determination to enforce arrangements, which, while protecting the demonstrators' freedom of speech and assembly, met our obligations under the Vienna Convention.[4]

In August, an official at the Foreign Office wrote to the Home Office on this matter:

> As you know, the City of London Anti-Apartheid Group (CLAAG) has been mounting a "vigil" since late April outside the South African Embassy in Trafalgar Square. The CLAAG has said that it will maintain the demonstration until Nelson Mandela is released. The group involved is small (normally less than a dozen people) but makes frequent and vociferous use of a loud-hailer. The South African Ambassador has complained many times since April to the Foreign Office about the noise (the demonstrators are on the pavement beneath the Ambassador's window).[5]

During July and early August 1986, the police attempted to utilize three different sets of regulations to prevent picketers from using their megaphone. They used breach of the Control of Pollution Act in order to restrict the use of megaphones at night. They charged a picketer under the Metropolitan Police Act (1839), claiming that she used "a noisy instrument to call people together" when she invited members of the public to join the Non-Stop Picket. They also started using the City of Westminster by-laws about 'noise pollution' to attempt to curtail the Picket's use of amplified sound. Technically, contravention of these bylaws was not an arrestable offence, and many picketers had their names and addresses taken by the police in order that they might be 'reported for summons'. Nevertheless, some picketers were arrested for 'noise pollution' during this period.

Faced with these attempts to curb the effectiveness of their protest, City Group and individual picketers responded in two ways. First, they stood their ground and contested the use of the noise pollution by-laws on the streets and in the courts – refusing to be intimidated out of protesting in the way they saw fit. Second, they worked behind the scenes to challenge (within Westminster City Council) how the by-laws were being applied. In July 1986, City Group had published a dossier cataloguing *100 days and nights: a record of police harassment*, which reported:

> The police have persistently attempted to stop the use of the megaphone and the embassy have complained about it. We have no wish to disturb residents in the area, especially at night, but we are also determined that our message is heard.[6]

On Monday 18 August 1986, 100 people gathered, gagged, outside South Africa House to pre-emptively break the silence that the Metropolitan Police were attempting to impose on the Non-Stop Picket. This attempt to prevent protesters from using their megaphone was just one of the ways in which the police attempted to curtail the Picket in its early months. Concurrently, there were also attempts to prevent the Picket from accepting donations from the public, as well as repeated harassment of black, female, and gay picketers, and a catalogue of violent arrests.

From 9 a.m. on that Monday, protesters gathered on the picket wearing gags to symbolize how the Metropolitan Police and the South African Embassy were attempting to silence anti-apartheid protest. At midday, in front of the gathered media, the protesters removed their gags and sang protest songs, before a succession of City Group activists and their supporters began making speeches using the megaphone. Speeches followed by Jo Richardson, Member of Parliament (MP), Sharon Atkins of the Labour Party Black Sections, David and Norma Kitson, Peter Tatchell, and the author Lynn Reid Banks.

The police immediately backed down, and no protesters were reported for summons. The senior police officer on the scene, Inspector Perry, even told journalists that he had no idea where City Group had got the idea the police were trying to silence them. Despite his assurances on this occasion, picketers' names and addresses were taken and reported for summons repeatedly that summer. On the picket, dealing with the police's use of the by-laws became quite playful at times:

> We played cat and mouse with [the police] – particularly over noise pollution. Once one person was warned for noise pollution, we would pass the megaphone up the line until the next person was warned. We knew that they had to warn each person three times individually before arrest. The police probably found us irritating, but they were protecting the Embassy of a criminal state.[7]

When a group of picketers was prepared to work together like this, they could continue using the megaphone for quite some time before the police came close to reporting individuals for summons or making arrests (although this was always a gamble).

With these tactics being used on the streets, leading members of City Group conducted quieter political work on the issue behind closed doors. Even so, the focus was always on undermining attempts to curb the Picket's noise. In October 1986, Andy Higginbottom, the Secretary of City Group, wrote to Andrew Dismore (then the leader of the Labour Group on Westminster City Council[8]) asking him to intervene around the issue:

> Since 18 August the police have taken individuals' names and addresses over 40 times during office hours 9 a.m. to 5 p.m.

He continued by requesting that the Labour Group would

agree a suitable weekday when councillors would again join us on the picket to test the police response.

While it is unclear whether the councillors did join the Picket to make a noise, in the way that was suggested, they did raise questions within the council chamber, challenging the political use of the noise pollution by-laws. The Labour Group on Westminster City Council responded by calling an extraordinary council meeting on 24 November 1986 to discuss concrete ways in which the Council could oppose apartheid (including material support for the Non-Stop Picket)[9].

City Group's resilience in defending their use of a megaphone paid off. Lorna Reid was arrested and charged under the by-laws on 6 October 1986. Her case became a test case for the use of the by-laws. She was convicted at her first trial but won a subsequent appeal when the judge and two magistrates ruled that the megaphone was not a noisy instrument compared to the background noise of vehicular traffic around Trafalgar Square. Following this judgement, the police intervention around 'noise pollution' tailed off (although the use of the megaphone remained contested for much of the Non-Stop Picket's duration).

The way in which the Non-Stop Picket responded to the attempts to curb the noise it made was repeated when the police (and the Embassy) contested other aspects of its practices over the years. The winter of 1986/1987 was a cold one. The pavement outside the South African Embassy was very exposed to the elements – the wind, in particular, seemed to gather force as it crossed Trafalgar Square. Anti-apartheid protesters standing on that pavement had no source of shelter from the wind, rain, and cold temperatures, especially at night. City Group was inventive in thinking about solutions to this problem. With images of industrial picket lines in their minds, on 12 January 1987, they decided to try to install a coal-fired brazier on the picket. The brazier was removed by the police. In response, City Group organised to ensure that a high-profile supporter, the Mayor of Islington, was there to witness their next attempt at lighting a brazier and so were press photographers (Figure 5.1).[10]

As one retired police officer remembered, the police deliberately tried to restrict any infrastructures that would make picket life more comfortable and more permanent. She suggested that, at least between the picketers and the rank and file officers who actually stood on point outside the Embassy, this frequently degenerated into a 'game':

> Initially the main points of contention were the Picket's continuous presence, was it obstructing the highway, tried and failed! Then the use of the megaphone, music and constant noise to disrupt those inside the building. The Picket would try to creep out and take over more of the footway, the use of brazier as mentioned above. We would challenge anything that made it easier or more comfortable for them to stay there, to

108   *Defending the right to protest*

JANUARY 1987 – THE COLDEST WINTER FOR FORTY YEARS Metropolitan Police extinguish the picket brazier, lit by the Mayor of Islington, at 2.30am before returning to their heated van
Ion the Non Stop Picket to release Nelson Mandela outside SA House, London. City of London Anti-Apartheid Group

Figure 5.1 City Group postcard recording Councillor Bob Crossman, Mayor of Islington, lighting the brazier on the Non-Stop Picket, January 1987. Source: City Group.

try and wear down their resolve but like us they just did their bit in shifts so it wasn't quite so uncomfortable. The arrival of chairs, tables, sleeping bags, tents or tarpaulins – it was all up for dispute. They'd bring along something new, we'd baulk at it – games.[11]

These comments offer some confirmation of the Metropolitan Police's strategy for containing the Non-Stop Picket by reducing its facilities and the space it occupied to the bare minimum. It also gives a fascinating insight into how some of the rank and file officers who policed the protest understood their role (and attempted to pass the time there).

In December 1987, staff at the Embassy wrote to the Foreign and Commonwealth Office twice in one day to complain about the activities of City Group.[12] In one of these letters, the Ambassador asserted that the Non-Stop Picket constituted an "impairment of the dignity of the Embassy." In choosing this wording, he was accusing the British government of failing in its duties under the Vienna Convention (1961) to protect the 'peace and dignity' of a diplomatic mission (Mamadouh et al. 2015). Of course, the protesters he complained about would have been unworried by this accusation – for them, the Embassy represented an illegitimate regime that had been condemned by the United Nations as a crime against humanity and, hence, deserved no 'peace and dignity'. Frequently, the Embassy's complaints to the Foreign and Commonwealth Office appear to have been followed a short while later by a renewed offensive by the police against the Picket (but this was not always the case). For much of its duration, the presence of the Non-Stop Picket and the methods it used to advance its protests were fiercely contested by the Metropolitan Police and other agencies, under political pressure from Whitehall, and diplomatic pressure from the South African Ambassador and his staff. The police and the picketers were engaged in a war of attrition and City Group deployed creativity, cunning, and civil disobedience to maintain and expand its presence outside South Africa House.

Relations with the police

While the picketers had chosen to spend long hours of their time standing outside South Africa House, for many of the police officers, this was not the way they expected to be repeatedly spending their working hours. In addition to the officially directed attempts to dislodge and undo the continuity of the Picket, boredom and antipathy often led individual officers to try and provoke arrests. The picketers, in turn, were not averse to winding up the police when the opportunity arose.

Former activists from the Non-Stop Picket remember their interactions with the police in a number of ways. Some, like Nicki, suggested that the relationship was largely shaped by indifference and (asymmetrical) disinterest: "We tried to ignore them but they didn't want to ignore us."[13] Others acknowledged that the relationship was more hostile and antagonistic, and that the hostility came from both sides (even as it was played out through unequal power relations):

> Utter hostility and contempt underlay all our contact with the police. Barely concealed sometimes. Norma [Kitson]: "Stupid pissholes!" muttered loudly under her breath as she walked away from an encounter.[14]

Penelope acknowledges that this hostility was frequently expressed by senior figures in the Group's leadership. This view was shared by several of the retired police officers that we interviewed:

> Relations were hostile, partly because some protesters were just 'anti-police' in general. But the fundamental point, I think, was the frustration of the protesters – i.e. the Police were protecting the Embassy and, by extension, the Apartheid Regime (as far as the protesters were concerned). That causes a public/private conflict for individual police officers.[15]

For many of the police officers, the Picket's rule about only engaging with the police through the designated Chief Steward on a given shift was particularly disconcerting:

I have been involved in public order policing on many occasions since, and was heavily engaged in anti-road building demos a few years later, I would say that a big difference was that there was always some more engagement with demonstrators, good and bad, that seldom happened at SA House.[16]

One senior officer remarked that, for a long running protest, it was difficult to establish any rapport with the protesters. In hindsight, many of the police officers expressed regret that they could not establish friendlier relations with the picketers. While it is likely that some (broadly sympathetic) individual officers might have preferred an easier mode of engaging with the protesters standing opposite them, it seems likely that the practices established by the Picket for interacting with the police did help destabilize the normal balance of power enough to give the protesters some tactical advantage at times. This was not accidental. As a result of their engagement in solidarity work with Irish Republicans, leading members of the Revolutionary Communist Group (RCG) had analysed the counter-insurgency strategy of Major General Frank Kitson (1971), who advocated low-intensity intelligence gathering to develop a total picture of dynamics within insurgent communities (RCG 1984). With this lesson in mind, alongside Norma Kitson's experience of covert operations for the African National Congress (ANC) underground in the 1960s, City Group's approach of limiting conversations with the police, and channelling them through experienced Chief Stewards, was a deliberate defensive strategy:

> I think we pushed them to the limit really. I think they didn't know how to deal with us and I think most of the things we did were quite, I think we were quite intelligent, quite clever in the way we presented ourselves, made our demonstrations, and obviously were always very diplomatic if ever spoken to by the police anywhere near the Embassy. I did have several confrontations away from the Embassy.[17]

In addition to the street-level interactions between protesters and police officers that took place in and around Trafalgar Square, there were also frequent discussions and negotiations between senior police officers at Cannon Row Police Station and City Group's leadership (particularly Carol Brickley). If particular difficulties arose on the picket, Carol would often phone senior officers at Cannon Row for an explanation. She would also negotiate with the police ahead of major demonstrations. David Gilbertson, a Chief Inspector at Cannon Row at the time, remembers:

> Relations overall were good. Contact at a senior level between the organisers and senior police officers was business-like, if somewhat 'frosty' (on both sides).[18]

From comparing the interview responses we can conclude that Gilbertson was probably more sympathetic to the picketers' cause than most of his colleagues. A Chief Superintendent from that time was more critical:

> The leaders would not compromise or negotiate at all. Their form of democracy was all they were interested in. They had no regard for other members of the public in the vicinity of South African Embassy, male, female, or children they insisted on doing what they wanted to do without regard to public safety.[19]

These differing opinions are significant. Once the Non-Stop Picket was fully established and had survived its first few months intact, and certainly once the right to protest in front of the Embassy had been regained in July 1987, the approach of senior officers at Cannon Row could make a significant difference to how the Picket was policed on the ground. There *were* periods of time (in the second half of the Picket's existence) when there were few(er) arrests unless City Group activists chose to make themselves liable for arrest through acts of direct action. Carol Brickley acknowledged that her style of negotiating was often to block, disrupt, or subvert the wishes of the police:

> We didn't do deals; we always really obstructed what the police wanted. Usually they wanted us not to have [an event] or to tone it down or whatever. I was quite good at handling those situations where we wanted to do something that the police didn't want us to do, and we would go ahead and do it.[20]

City Group had a political objective to disrupt the work of the South African Embassy and draw attention to the ways British political and economic links with the apartheid regime were "distributed along long chains of command" (Massey 2008: 323). They framed any attempt to curb their actions (and their attempts to disturb the Embassy) as a political attempt to curb a legitimate protest. Dominic's view of the relationship with the police encapsulates this view:

> There probably were moments of confrontation, but really the hostility was from the police to the Picket. The Picket was there to disrupt the working day of the South African Embassy and the cops were hostile, and so we all ended up getting arrested lots of times. And some of those times would have been because we were making a political point and we weren't going to back down.[21]

Despite this frequently tense, if not hostile, relationship, both picketers and police officers remembered moments when the picketers responded to unexpected events and temporarily dropped their hostility to police officers as representatives of a state institution and responded to them as individual human beings instead. Eleanor Cave remembered a picketer "jumping in to help when a police [officer] had a seizure."[22] Similarly, a former woman police officer from Cannon Row recalled the following event on the picket:

> Weirdly one of my most abiding memories was of being on the picket one evening and a call for assistance from a colleague coming over the radio. There was a fight happening around the corner in the Strand. I remember leaving my post and running round to help out, having a bit of a roll around on the floor helping to arrest a drunken yob and then having to trot back to the Picket as by right I shouldn't have left it in the first place but some things would always take precedence. I was obviously out of breath, a bit pale and the after effects of the adrenaline had kicked in and my hands were shaking. Steven Kitson was on the picket that evening and after looking at me in a concerned fashion for a minute or two he came over and asked me if I was alright. I was really rather touched. You have to appreciate there was very little contact with the pickets, they didn't talk to us and we didn't to them unless it was to raise an issue. It was a very nice gesture.[23]

As this officer acknowledged, such moments of concern and care (in either direction) were seldom publicly expressed.

Crossing the road

City Group activists were fond of paint (and paint stripper[24]). In the pages that follow, we analyse the events that occurred as a result of three young activists throwing gallons of red paint over the main entrance to the Embassy in May 1987. The paint-throwing action on 6 May 1987 was one of a small number of open and overt actions of this kind. More often, the Embassy, its cars, and other property associated with the apartheid regime's representatives in London were targeted covertly under cover of darkness. On these occasions, the actions were often claimed by the pseudonymous 'Chuck Paint' of 'Red Reprisal'.[25] These minor acts of sabotage might have been little more than an inconvenience to the Embassy, but they boosted the morale of City Group supporters.

The overt action against the Embassy on 6 May 1987 was a protest against the white-only general election taking place in South Africa on that day. This action functioned in several registers. Visually, the red paint covering the embassy doors was highly symbolic of the blood spilled by the apartheid regime. In this respect, the action served to bear witness to the violence of apartheid. The action also served to disrupt the normal functioning of the Embassy (including its use as a polling station, that day, for ex-pat South Africans in Britain). Finally, the action (including the subsequent trial) was a publicity stunt that provided an opportunity to educate the public about the reality of life under apartheid and the lengths to which the apartheid

Defending the right to protest 113

regime was prepared to go to silence anti-apartheid opposition (both within and beyond its national borders) (Israel 1998; Bell with Ntsebeza 2003).

Throughout the first year of the Non-Stop Picket, South African diplomats had exerted pressure on the British government to restrict and ban the protest outside their front door. The paint-throwing incident provided the police with the excuse they needed in order to act decisively. The police used an arcane Victorian by-law, "Commissioner's Directions,"[26] which allowed the Metropolitan Police Commissioner to curtail public gatherings within a mile of Parliament, to allow MPs free movement to go about their business, to ban the Picket from outside the front of South Africa House. The Picket relocated to the steps of nearby St Martin-in-the-Fields Church, on the opposite side of Duncannon Street from the north side of the Embassy. They would remain there until early July, while the ban remained in force. Nevertheless, City Group and their supporters did not accept the ban without a fight. From the first evening that the ban was imposed (and throughout the next two months), activists repeatedly risked arrest to break the police ban on their protest and defend the right to protest outside the Embassy.

In hindsight, it seems clear that the Metropolitan Police were seeking an excuse to use this power to restrict the Non-Stop Picket. A week earlier, on 30 April 1987, they had temporarily applied the Commissioner's Directions by-law to force the Picket to move from its normal spot directly in front of the embassy gates to the corner of Duncannon Street and Trafalgar Square (away from all entrances to the Embassy). On that occasion, they used the Territorial Support Group (the Metropolitan Police's specialist public order squad) to forcibly move the protest. In the process, they violently arrested two picketers – Lorna and Dele. Lorna was charged under Section 14 of the (then) new Public Order Act, becoming the first person in the country to be charged under this section of the Act. When their case eventually came to court, the magistrate (having seen the police video of the incident) dismissed the charges without even hearing the defence evidence and declared that the Public Order Act did not negate the right to peaceful protest (Bailey and Taylor 2009).

Direct action precipitates police ban

The idea to protest against the white-only election in South Africa by throwing red paint over the Embassy came from Irene – at 16, the youngest of the three activists involved. She shared the idea for the protest with her sister, Liz, who was a couple of years older and also involved with the Picket. She then approached a third picketer but did not pursue the plan with him:

> And I actually, for whatever reason, I can't remember, but I think I broached the idea originally with him and said, oh what do you think about doing it, and he'd said, yeah, I want to do it with you, and then I bottled it because I just didn't quite trust him enough, so I asked my

sister and I asked Adam. I don't know why I asked Adam, he was a friend and he was quite feisty I suppose.[27]

Adam appreciated the potentially serious charges that the trio could face as a result of their actions but agreed to take part because, "it was a significant day, so I did feel that it warranted a fairly dramatic response."[28] The trio discussed their plans, in very general terms, with a leading member of City Group:

> I don't even remember it as me going to [them] for authority. Maybe I was just saying, maybe it was just that I was lacking in confidence to do it and I was saying, what do you think? I don't think it was a thing of, oh I better talk to someone on the committee, I don't think there was a formality. I don't think it would have occurred to me at that age that I should ask anyone to do anything![29]

With no explicit attempt having been made to dissuade them, they made more concrete plans for their action on the day of the election. This included tipping off a sympathetic photographer:

> I think I ... spoke to him a couple of days before and said, could you come and, could you be there, and I didn't tell him what we were doing. So I definitely had a sense of this is a top-secret affair.... I'd said to him, can you just be there and have a camera at whatever time of day, I think it was 10 a.m. or something. He had no idea what we were doing, I don't think. I don't think he had an idea what he was doing, because I think I was quite conscious that because there was this feeling that there might be informants, for want of a better word, that you just wouldn't say anything you didn't need to say, and I was conscious that if the police knew we were going to do it that they would be able to stop it quite easily just by putting barricades up and then you're screwed immediately.[30]

On the day of the election, they met as planned but quickly realized that they had not discussed exactly what they would do next. As Irene recalled:

> I think we had a tin [of paint] each. I remember we met in Charing Cross Station and we took the lids off in Charing Cross Station and put them in a dustbin, and then walking up the stairs as a group of three and then we had no idea what we were doing, that's my memory, as in it wasn't like, oh you go there and I'll go there, it was just from that point there was no organisation at all. Well, we'd just not planned it at all. We'd planned where to meet, to have the paint, and then we got there and started marching towards the Embassy and you think, well, I can't go back now because the police could probably see us at that point, so you had to just keep going forward, and I remember going like this with the paint and nothing happening! Because I was quite small and it was quite

thick, and I went to throw it at the Embassy and literally nothing happened, it might have been like a splodge on the floor and just thinking, this is not working, and so I ended up doing it with my hands because I couldn't, I didn't have the power in my arms to throw it![31]

Adam remembers that moment slightly differently:

> I think we were a bit surprised by how effective it was to be honest ... We were able to throw all the paint. We had plenty of time. So we had quite a lot of paint and we were able to, you know, if you've seen the pictures it is quite impressive, which is surprising really. And because of the nature of the paint it fucked up the electronic doors, they couldn't open the electronic doors. You know that glass door as you go through the alcove thing? They couldn't open that.[32]

Although there were police on duty in front of the Embassy, and one got covered in some of the paint, Adam remembers that they were taken by surprise and seemed slow to react.

> The Old Bill were complete numpties. We could have walked away; we could have got away; we could have been about five miles away by the time they got their act together. [Interviewer: So why didn't you?] Well I think because we knew it was more than just actually about throwing the paint. You know, it was about actually saying well this is what we've done, what are you going to do about it?[33]

Irene also acknowledged that they could have run at that point, but that did not occur to her. Her attention was on the reactions of their fellow protesters on the picket. Although it was an election day in South Africa, and there was a rally planned for that evening, at 10 a.m., there were relatively few people on the picket:

> So they all went crazy, it was really good! They were really cheering and I think there's a picture in there of us getting arrested and it's just the faces of the people on the picket just curled up with laughter. It must have just seemed like such a shock that you're standing there at 10 a.m. and then this thing happens in front of you, and there was one, I think there was one policeman on duty and it was his like second week at work, he'd just qualified. There might have been two but they were both, they weren't the hostile ones, they weren't the ones who'd come on and taunt you, they were the ones that turned up and stared at their feet! And I just remember them being completely bamboozled by it, trying to run around and decide who to arrest. There were three of us and two of them or maybe even one of them at that point. I think it might have only been one actually, the young lad ... we just kept smearing paint and the picketers were chanting and laughing and shouting.[34]

Amanda was serving as the Chief Steward on the picket that morning:

> I had an idea that it was going to happen, but yeah, I was informed that something may happen. I wasn't 100 per cent; I didn't know the details of it. But that's what's so funny because it's quite random, because I could have been standing there quite innocently as a Steward whatever, and a certain group of people come down and – take their action, and it could have been a toss of the coin whether I could have got arrested alongside them or not.[35]

In hindsight, Irene accepts that the action was not entirely successful:

> I don't think it ever occurred to me I [could] go to prison ... I think I just wanted to do it. I think there was a big build up in the media and obviously on the picket in City AA, there was a big build up towards these white elections that it felt like a moment in history, if you like, that they were going to have these elections and I just felt like we needed to do something about it and that was what I came up with, was the – ... idea. Well, I think it was based on the idea, obviously red representing blood spilled by apartheid, but also an intention to close the Embassy which completely failed and was naïve possibly to think it would, but – the idea was if you're not allowing black people to vote then we'll stop everyone voting and so that was the idea is close the Embassy through making it impossible for people to walk through the door, but obviously they just went through the side entrance! So I'm sure it was completely, in practical terms, meaningless for what we were trying to achieve, but obviously the things that happened from it perhaps had more significance.[36]

While the junior police on duty that morning might have been slow to react at first, before too long, their superiors had authorized a more thorough response – within hours, they forced the Non-Stop Picket to relocate across Duncannon Street and banned protests directly in front of the embassy building. The rally planned for that evening was loud and militant, but it took a very different course to that which had been planned, as picketers began their campaign of civil disobedience to win back the right to protest in front of the Embassy.

We asked Irene if she ever felt guilty that this was one of the consequences of her actions that day. She replied:

> No, I don't think I felt bad, no, because it carried on and it gave, what's the word, not purpose, but it gave an additional kind of – fight for, yeah, so I don't think I ever felt guilty about it.[37]

Irene herself would be arrested another three times that summer, defying the ban on the Non-Stop Picket.

Defying the ban

On the evening of 6 May 1987, London anti-apartheid activists should have been rallying to protest against the whites-only election in South Africa that day. They gathered that evening with that intention, but their attention ended up being focussed closer to home. Having fought hard over the previous year to maintain a non-stop picket, in the face of dozens of arrests, petty harassment, and numerous court cases, City Group was not going to relinquish the site outside the embassy gates without a fight. They responded swiftly to this latest attack on their right to protest, drawing on their experience of successfully defeating a previous ban in 1984 through the South African Embassy Picket Campaign (see Chapter 2).

That evening, following Norma Kitson's lead, they decided to contest the ban with direct action. Twenty picketers (including most of City Group's leadership) crossed Duncannon Street, defying the ban, and attempted to continue their protest outside the embassy gates. They were arrested. The following day, a further six activists crossed the road. Over the following eight weeks, at least 151 picketers and their supporters defied the ban, resulting in a total of at least 172 arrests.

The initial wave of civil disobedience to break the police ban was taken by leading members of City Group – Carol Brickley, the Group's Convenor; Norma Kitson, the Deputy Convenor; and Steve Kitson. They were joined by Adrian States, a Labour Councillor for the London Borough of Camden. The case against these four protesters would later become one of the legal test cases in challenging the police ban in the courts. Adrian States was the vice chair of the Camden Council's police sub-committee at the time, so his arrest under the new and controversial Public Order Act attracted media attention. According to a report in the *Camden New Journal*, Camden Labour Party activists heard of his arrest during the annual general meeting of their local government committee, which they promptly interrupted in order to attend Cannon Row Police Station and demand his release[38].

The campaign of civil disobedience continued each Friday during the Group's scheduled weekly rally, but also – less predictably – at other times during the week to maintain an element of surprise. One of those who was inspired to join the campaign was Penelope. She recalled how it was Norma Kitson, specifically, who convinced her of the importance of defying the ban, sharing an extract from her diary at the time with us:

> The Crossing the Road campaign was Norma's masterpiece. [Here's an] extract from my journal June 3, 1987 just after the Picket had been moved:
> Last Sunday I was mobilised, you might say I was put a firecracker under by Norma Kitson at a workshop at her house. She said that confidence was in the eye of the beholder. She said confidence was a DECISION. She said that if asked, we were to say precisely this: 'The police have banned

118   *Defending the right to protest*

> *the picket from the pavement'. Using THOSE words and NO others. She went on to talk about how to develop communication skills [when trying to mobilise others to join the campaign] – rule one – commitment and belief in your cause – rule 2 – don't communicate anxiety [don't say e.g. there's a crisis here, we need someone desperately]. She said we have lost our indignation in this country – it is not enough to tell them the facts, though knowing the facts is important – you have to communicate indignation. Do not be cool – be involved. [...]*
>
> *Later on there were a series of workshops for people who were prepared to get arrested in order to win the pavement back. "This is going to be a successful campaign. It is going to be very well organised. You will get arrested, but you will NOT get a criminal record. They have invoked commissioner's regulations, which are archaic and irrelevant and so this is low attempt to trick us and scare us which won't work because we have a good legal defence against it, which WILL work. The British people DO HAVE a right to demonstrate peacefully on the streets of their capital city. We will win our pavement back."*[39]

Penelope went on to recall:

> [Norma] explained it so well, communicating white-hot indignation and absolutely no anxiety, and gave us assurances about the legal and peaceful nature of this prime piece of civil disobedience ... Any time I show any signs of being supine in the face either of authority or of peer pressure I remember that, and that keeps me going.[40]

Inspired by Norma's confident indignation, many City Group activists, young and old, were prepared to risk arrest for the right to protest against apartheid. City Group was also successful in mobilising local councillors, prospective parliamentary candidates and trade-union activists from across the country to stand in protest with them. The longer their campaign continued, the more inventive and cunning the protesters had to be in order to get across the road to take their stand. Photographs in the archives show groups of women hopping off the open-back platform of a Routemaster bus outside the embassy gates and defiantly unfurling their banners.

After nearly two months, City Group won back the right to protest where they chose. Precisely because the Commissioner's Directions by-law existed to allow MPs to freely go about their parliamentary business, City Group found a novel way to use this law against the police in order to re-establish the Non-Stop Picket in front of the Embassy. On 2 July 1987, four Labour MPs joined in defying the police ban on anti-apartheid protest. Tony Banks, Dennis Canavan, Harry Cohen, and Allan Roberts, carrying City Group placards, lined up to protest outside the embassy gates. Television crews and press photographers were there in force, crowding around to witness their action. The senior police officer on the scene, Superintendent Little, chose not to arrest them. They were soon joined by Norma Kitson, David Yaffe, and other leading City Group activists. Superintendent Little was forced to explain that Commissioner's Directions did not apply to the MPs but did apply to the other protesters. Nevertheless, while the MPs were present, he declined to arrest the others. His compromise was to create a 'designated area' (a pen of police crowd control barriers placed to the north of the embassy doors) where City Group could protest temporarily. Norma and David remained directly outside the embassy gates protesting alongside the four MPs, while other picketers occupied the adjacent 'designated area'. Eventually, their work done, the MPs left the area, but David and

Figure 5.2 Norma Kitson and David Yaffe defy the police ban on the Non-Stop Picket, 2 July 1987, photographer unknown. Source: City Group.

Norma stayed put. Half an hour after the MPs' departure, Superintendent Little arrested David and Norma, under Commissioner's Directions, for doing exactly what he had allowed them to do while the MPs were present. But, by that time, it was too late. City Group had re-established their right to protest against apartheid where they chose – directly outside the South African Embassy and not across the road from it on the steps of a church (Figure 5.2).

City Group's campaign combined civil disobedience to break the ban and make it unworkable; political campaigning to raise the issue and question the ban in Parliament and in the media;[41] and legal challenges to the charges brought against those arrested during the defiance campaign. With the ban effectively overturned on 2 July, eventually all of the 172 charges brought against City Group supporters during the 'cross the road campaign' were dropped or thrown out of court.

Politically, City Group understood the police ban on their protest as an expression of British collaboration with apartheid. By making the ban unworkable, City Group activists believed that they were not only defending the right to protest in Britain but contesting the British Establishment's protection for apartheid's representatives in London to go about their business unchallenged. This political analysis was highlighted in the slogan used on the posters throughout the defiance campaign – "Botha's laws come to Britain." For many people associated with the Non-Stop Picket, extending solidarity to those resisting apartheid in South Africa could not be separated from challenging racism in Britain or resisting state attempts to curtail the right to protest. Their solidarity did not just flow in one direction – from the streets of Central London towards South Africa – but stemmed from an understanding that political change in Britain and South Africa were intimately linked. But, having sustained the Non-Stop Picket outside the embassy gates continuously for over a year, City Group members were fighting to regain control over a small patch of the city that they had made their own.

After a year of successes in deflecting attempts to restrict their protest, it is hardly surprising that picketers and their supporters reacted so angrily when the Metropolitan Police moved them under the Commissioner's Directions. That by-law was designed to facilitate the free movement of those bodies essential to the functioning of the liberal democratic state; but it was used to limit the presence of protesters' bodies that challenged its policies, authority, and legitimacy. For both sides in the dispute (as well as third parties such as the parish of St Martin-in-the-fields, who got drawn into the conflict[42]), there was a lot at stake.

In total, 172 people were arrested during City Group's campaign to break the police ban and defend the right to protest.[43] All charges were eventually thrown out of court. During the period of the ban on the Picket, City Group supporters acted in an unruly manner, breaking the law in order to defend their right to protest and thereby defend a space in which to be unruly (Brown 2013). The irony of both the manner in which the Picket was restored to its place in front of the Embassy and picketers' subsequent legal victories in court was that the Metropolitan Police were exposed as bending legal 'rules' in order to curtail unruliness.

Using the courts to expose apartheid

The defeat of the police ban on the Non-Stop Picket was not the end of the story that began with the paint-throwing protest. Adam, Liz, and Irene were charged with criminal damage to the embassy doors, a policeman's uniform, and his long johns. They used their subsequent trial(s) as a platform to expose the complicity of the South African Embassy in state-sanctioned violence against its opponents, and to gain greater publicity for their cause.

This trial, in particular, was fought politically. Adam, Irene, and Liz did not deny that they had thrown paint over the embassy entrance but they pleaded not guilty, arguing that they had taken their action to prevent the (greater) crimes of the apartheid regime's agents operating out of the Embassy. A number of expert witnesses were arranged to speak in their defence at the trial. David Leigh of *The Observer* gave evidence about the illegal activities of the Embassy and its staff, including their involvement in the bombing of the London offices of the ANC in 1982 (when explosives and detonators were smuggled into the UK in diplomatic bags) (Bell with Ntsebeza 2003; Ellis 2013). Zolile Keke, the Chief Representative of the Pan-Africanist Congress (PAC) in London, and Norma Kitson gave evidence about their experiences of torture in South Africa. These and other testimonies were enough to convince the jury. On 30 September 1987, the jury at Southwark Crown Court refused to convict them, despite an instruction to do so by the presiding judge.[44]

The case went to a retrial, and, on 23 March 1988, Adam Bowles was sentenced to 14 days in Brixton Prison for his part in throwing the paint over the doors of South Africa House.[45] At this second trial, Adam and his co-defendants once again claimed that they had committed the 'criminal damage' to the Embassy in order to prevent the commission of a greater crime – apartheid. In court, Adam made the following statement to this effect:

> The United Nations has proclaimed that the South African constitution is null and void and any elections within these territories is null and void ... What we are talking about is three people ... taking action to stop the phony, proxy, so-called elections[46].

Following the jury's guilty verdict, Judge Morland summed up the case by acknowledging that Adam, Liz, and Irene were "of the highest moral and political character."[47] Nevertheless, he fined both Liz and Irene, giving them two-year conditional discharges, and imprisoned Adam for two weeks in what he hoped would be a punitive deterrent. He served only a week in gaol and received cards from supporters all over the country. On the Saturday of his imprisonment, City Group supporters staged a noisy solidarity protest around the perimeter of Brixton Prison.[48] Adam was one of only a very small number of City Group activists who were jailed during the Non-Stop Picket.[49] They all received the type of solidarity and support that he did.

Although the Metropolitan Police tried many tactics to curtail the Non-Stop Picket, the Picket never had to endure the frequent eviction attempts that have characterized life at other long-term protest camps (Roseneil 2000; Feigenbaum et al. 2013). City Group always maintained that the Non-Stop Picket *was* non-stop from 19 April 1986 until Nelson Mandela was released from gaol. That is true; although, there were at least two occasions when the police did manage to briefly interrupt the continuity of the Picket. Once, in the early months of the Picket, in 1986, the police arrested all three picketers on duty and took them and all of the Picket's equipment into custody. A similar incident occurred in the final months of the Picket in 1989. On both occasions, the arrested picketers exercised their right to a phone call from custody and notified their comrades of what had happened. City Group sprang into action and quickly replaced the protest with new personnel and a banner.

Having defended the Picket against so many attempts to curtail it during its first year, regular picketers had developed a strong territorial attachment to 'their' few square metres of pavement in front of South Africa House (Leitner et al. 2008; Cumbers and Routledge 2013). As James Godfrey explained:

> There we were in this geographic space in which we conducted our business, and we conducted our business of protest against apartheid and the police conducted their business of protecting the rights of the South African Embassy under the Vienna Convention and everything else, and the peace and good order of whatever, the word, breach the peace and protecting upright citizens walking down the street. So it was over a limited patch, a contested space.[50]

The defeat of the police ban enabled City Group to claim with some legitimacy that they had maintained a continuous protest at the Embassy until Mandela was released – the few short disruptions to the non-stop nature of the protest were a result of the actions of the Metropolitan Police, not the picketers' neglect of their duties (and this gave their solidarity additional meaning). The intensity of the campaign to defeat the ban made it one of the most memorable moments in the Picket's history for many who were involved at the time (whether they took part in civil disobedience to defy the ban or not). The confidence instilled by this success had long-lasting effects – in November 1989, officers approached Lorna Reid on the picket and told her to move back (behind the line of placards on the ground). She confidently replied that she would stand on the same line she had since April 1986.[51] By that stage in the Picket's history, after 3 and a half years of non-stop protest, City Group activists were confident in determining for themselves where the Picket should stand. It is for these reasons that we choose to examine the events surrounding the ban in such detail – in several different ways, they exemplify City Group's attitude to the police, to the law, and their practice of taking direct action against apartheid.

Notes

1. J. Kempster (1988), 'Report on the Non-Stop Picket of South Africa House' (excerpt from *Black Londoners*, 18 April), *BBC Radio London*. A recording of this report is deposited in the British Library National Sound Archive, C1499/1 C2. BBC copyright content reproduced courtesy of the British Broadcasting Corporation. All rights reserved.
2. Interview with Carol Brickley.
3. Email correspondence with retired Deputy Assistant Commissioner David Gilbertson. David Gilbertson was posted to Cannon Row Police Station as a Chief Inspector in 1986. In September 1987, he became the Superintendent in charge of the Central London Territorial Support Group. He was seconded to South Africa in 1993–1994 to assist with preparations for the post-apartheid elections. He retired from the Metropolitan Police as Deputy Assistant Commissioner in 2001.
4. Excerpt of a minute from – (FCO official) to Private Secretary dated 28 April 1986. Foreign and Commonwealth Office (n.d.), 'City of London Anti-Apartheid Group (CLAAG) – Activities/Non-Stop Picket of South African Embassy, London 1985–91'. Response to FOI Data Access Request.
5. Excerpt from a letter from – (FCO official) to – (Home Office official) dated 7 August 1986. Foreign and Commonwealth Office (n.d.), 'City of London Anti-Apartheid Group (CLAAG) – Activities/Non-Stop Picket of South African Embassy, London 1985–91'. Response to FOI Data Access Request.
6. City of London Anti-Apartheid Group (1986), *100 Days and Nights: A Record of Police Harassment* (28 July), p. 3.
7. Interview with Andy Privett.
8. From 1997 until 2010, Dismore was the Labour MP for Hendon. At present, he is a member of the London Assembly, representing Barnet and Camden.
9. Letter from Andy Higginbottom to Councillor Andrew Dismore, Westminster City Council, 18 October 1986; letter from Andrew Dismore to Andy Higginbottom, 26 October 1986; letter on behalf of Andrew Dismore to City Group, 11 December 1986.
10. Handwritten 'report on the brazier' by Richard Roques presented to the City Group committee, 26 January 1987.
11. Email correspondence with anonymous retired woman Police Constable based at Cannon Row Police Station in 1980s. This officer joined the Metropolitan Police when she was 18, in 1980. Cannon Row was her first post after completing her training. She served there until June 1987. Like many junior officers, she was frequently allocated to policing the Picket.
12. Foreign and Commonwealth Office (n.d.), 'City of London Anti-Apartheid Group (CLAAG) – Activities/Non-Stop Picket of South African Embassy, London 1985–91'. Response to FOI Data Access Request.
13. Interview with Nicki.

14 Interview with Penelope Reynolds.
15 Email correspondence with an anonymous retired Police Constable who was based at Kensington police station in the mid-1980s. He was drafted in to assist with policing the first anniversary of the Non-Stop Picket in April 1987.
16 Email correspondence with Nick Westwood, who was a police constable at Rochester Row Police Station in the early 1980s. He joined the Metropolitan Police in 1983. From 1984 to 1988 he was a beat officer based at Rochester Row Police in Victoria. He would often be allocated night shifts on point at the South African Embassy.
17 Interview with Simon Murray.
18 Email correspondence with David Gilbertson.
19 Email correspondence with an anonymous retired Metropolitan Police Chief Superintendent. He was a career police officer who joined the Metropolitan Police in 1959. From 1984 until shortly before he retired in 1989 he served at Cannon Row Police Station.
20 Interview with Carol Brickley.
21 Interview with Dominic Thackray.
22 Interview with Eleanor Cave.
23 Email correspondence with retired woman Police Constable from Cannon Row Police Station.
24 In an article in *The Telegraph* after plans for Mandela's release had been announced, Louis Mullinder, a counsellor at the embassy, said (about the Non-Stop Picket): "I remember it was 1,000 days last January because that's when my car got sprayed with paint stripper". Fletcher, K. (1990), 'Picket squares up to change', *The Sunday Telegraph*, 4 February.
25 These covert actions did not tend to be publicised in *Non-Stop Against Apartheid*, City Group's outwards-facing newsletter; but they did get reported in the Group's weekly internal bulletin, *Picketer's News*. For example, 'Paint Job', *Picketer's News*, 27 May 1988, reported that 'Chuck Paint' had called the Group's office to report that 'Don't Fly Apartheid' had been painted on the front of the South African Airways offices in Oxford Circus. A few weeks later, 'Blood on his hands', *Picketer's News*, 8 July 1988 reported on slogans sprayed across the front of Chatham House that week, before Jonas Savimbi of UNITA, South Africa's Angolan allies, was due to speak at the Royal Institute of Foreign Affairs. *Picketer's News* on 22 July 1988 reported that a City Group member, Naomi Freeman, had been arrested for throwing red paint over the South African Airways offices in Oxford Circus on 19 July. Reportedly, Naomi issued a statement saying, "I used red paint because I couldn't afford a bazooka."
26 An overview of the law regarding 'Commissioner's Directions' is provided by Louise Christian (1987).
27 Interview with Irene Minczer.
28 Interview with Adam Bowles.
29 Interview with Irene Minczer.
30 Ibid.
31 Ibid.
32 Interview with Adam Bowles.
33 Ibid.
34 Interview with Irene Minczer.
35 Interview with Amanda Collins.
36 Interview with Irene Minczer.
37 Ibid.
38 'Councillor arrested under new demo law', *Camden New Journal*, 14 May 1987. This press cutting was sent to City Group by Councillor Adrian States on 16 May 1987.

39 Interview with Penelope Reynolds.
40 Ibid.
41 Hansard HC Deb, 9 July 1987 vol. 119 cc211-2W; letters from Councillor Andrew Dismore of Westminster City Council to Commander Marnoch and Sir Kenneth Newman of the Metropolitan Police opposing the banning of the Non-Stop Picket from outside the embassy, 2 June 1987.
42 Letter from Richard Roques, on behalf of City Group, to Bishop Trevor Huddleston (25 May 1987) asking him to speak to his 'personal friend' the vicar of St Martin-in-the-fields Church. The letter states: "We are concerned that the vicar is being fed information by those who are hostile to us. The police visit him regularly and he has indicated that he may ask them to move us."
43 City Group's own publications variously give the figure for the number of arrests as between 169 or 172 (and the number of separate arrestees as anything between 151 and 169). We take the figure of 172 from a Written Answer provided by the Home Secretary, Douglas Hogg to Tony Banks MP on 9 July 1987 (Hansard HC Deb vol. 119 col. 211-2W, 9 July 1987).
44 City Group (1987), 'Jury refuses to convict anti-apartheid paint throwers', *Non-Stop Against Apartheid*, 23 (October), p. 2.
45 C. John (1988), 'Adam Bowles – Imprisoned – for fighting apartheid', *Non-Stop Against Apartheid*, 27 (April), p. 7; A. Bowles (1988), '7 days in Brixton Prison', *Non-Stop Against Apartheid*, 27 (April), p. 7.
46 A. Bowles (1988), '7 days in Brixton Prison'.
47 C. John (1988), 'Adam Bowles – Imprisoned – for fighting apartheid'.
48 M. Burgess (1988), 'Picket of Brixton Prison', *Non-Stop Against Apartheid*, 27 (April), p. 6.
49 We think only three picketers were gaoled as a direct result of a conviction for their anti-apartheid campaigning during the 1986–90 Non-Stop Picket. However, we are also aware of a handful of instances where homeless activists were remanded following arrests on or near the protest.
50 Interview with James Godfrey.
51 M. Farmaner (1990), *Record of Police Harassment on and around the NSP, 28th September 1989–4th January 1990*, p. 12.

6 Being unruly

Both on and off the Non-Stop Picket, City Group encouraged and enabled direct action against the representatives of the apartheid regime (and their supporters) in Britain. This chapter examines how picketers learned to be unruly in different contexts and various embodied ways. Through their non-violent, but confrontational political stance, the young picketers learned to think and act against the (British) state. Consequently, we examine the practices through which City Group offered political and legal support to those arrested on its protests. Their approach was particularly effective – of the more than 700 arrests associated with the Non-Stop Picket, over 90% of cases were (eventually) won by the defendants. In contrast to many contemporary social movements, City Group was quite hierarchically organised. This had advantages in terms of legal defence; but it was also sometimes in tension with how some picketers interpreted and embodied the tactical unruliness fostered by the Picket.

The culture of the Picket not only conveyed its political message of solidarity, but helped individual participants define their personal identities (Thörn 2009). For many picketers, but particularly the young, the opportunity to stand on a pavement in the centre of London singing, shouting, and pushing the limits of legality with the police was powerful and empowering. To be on the picket was to bend or break the rules of appropriate behaviour in public space (Brown 2013). This chapter examines both the positive, empowering aspects of being unruly in relation to the state; but also considers the understandable tensions that this constant, low-intensity conflict could lead to. In doing so, we highlight the uneven terrain of this unruliness, demonstrating how some picketers could take (and get away with) the risks associated with being unruly more safely than others (see Chapter 9).

Actions on the picket were undertaken for a number of reasons. Sometimes they sought to directly disrupt the functioning of apartheid's representatives in Britain; sometimes they were designed to produce photo opportunities or significant arrests that could generate press coverage for the Picket and its message; and, at yet other times, they were more symbolic. For example, on

11 October 1986, during a protest calling for the release of political prisoners in South Africa called 'break the chains', three picketers chained themselves to the gates of the Embassy. One of them, Sally, recalled:

> I chained myself to the gates of Embassy. I was charged and convicted of Highway Obstruction. Richard Branson's father was the magistrate and he summed up by saying that he understood where we were coming from as he had fought against fascism in the war but he said on this occasion we had over stepped the mark![1]

On another occasion, in August 1988, a group of women activists marked South African Women's Day by taking scrubbing brushes to the fabric of the Embassy to wash off the bloody taint of apartheid and draw attention to the double oppression of women under apartheid (Bozzoli with Nkotsoe 1991; Lee 2009).

These actions embodied the Picket's spirit of unruliness – part of the challenge (and fun) of even the most low-key action was to see how far the police could be pushed. The unruly presence of the Picket was often performed through small acts of defiance that sought to assert and expand the space controlled by the protesters outside the Embassy. In addition to using direct action to defend (and consolidate) space for the Picket and its chosen practices (see Chapter 5), City Group's young activists also used direct action tactics in pursuit of many other campaigning goals. Thus, our discussion of unruliness on the picket is complemented by accounts of a series of occupations of the offices of South African Airways (SAA), and actions taken in pursuit of the sports boycott of South Africa. Whether active participants in these actions, or witnesses to them, picketers learned to take an unruly stance in relation to the police, the courts, and other representatives of the British state.

Given City Group's tendency to use direct action tactics to protest Britain's links with the apartheid regime in South Africa, and their use of intransigence, cunning, and civil disobedience to challenge attempts to restrict their protest, relationships between the Non-Stop Picket and the Metropolitan Police could be tense. Many arrests resulted. The archive of City Group's papers does not help reach a definitive figure on the number of arrests that occurred as a result of the Non-Stop Picket. During the Picket's second anniversary rally in April 1988, both Carol Brickley and Norma Kitson mentioned a figure of 600 arrests in the previous two years. By the end of the Picket, a figure closer to 800 was often quoted (although we can find little direct corroborative evidence of this). During the course of our research, we submitted a number of data requests to the Metropolitan Police Service under the Freedom of Information Act; the information that they eventually released, after much chasing, was extremely partial and did not aid in our assessment. Our request asked: "How many people were arrested in the vicinity of the South African Embassy, in relation to anti-apartheid protests, between 1 April 1986 and 24 February 1990?" and a number of subsidiary questions. We received the following reply:

> Data only exists covering the period April 1987 to February 1990. Details were only recorded of those who were arrested and charged. Where an individual was arrested and charged, information was not always available to denote whether this resulted in a conviction.
>
> The data is as follows: Between the period April 1987 to February 1990, 64 arrests were made in the vicinity of the South African Embassy. As some people were arrested more than once, this amounted to 54 people.
>
> A) Of those arrested, all were charged.
> B) The breakdown of the charges were as follows:
>
>> ABH on police
>> Assault with intent to resist arrest
>> Assault on police
>> Assault occasioning
>> ABH
>> Affray
>> Breach of probation
>> Criminal damage
>> Disorderly behaviour
>> Highway obstruction
>> Obstructing police
>> Public order-causing harassment, alarm or distress
>> Public order-fear of provocation or violence
>> Street trading
>> Threatening behaviour
>> Using insulting words and disorderly behaviour.
>
> C) Of those arrested, from the information available 34 were convicted of the above offences".[2]

There appears to be a significant discrepancy between the data that the police were prepared to release (a quarter of a century later) and the statistics frequently cited by City Group activists at the time. Some of this might be accounted for by the number of people arrested and released without charge (as these, it would seem, were not recorded in the police statistics). Certainly, in the months following the start of the police record (April 1987 onwards), City Group's papers suggest a far higher rate of arrests. From 6 May to 2 July 1987, the Picket was banned under Commissioner's Directions from taking place directly in front of the Embassy. During that two-month period, up to 172 arrests were made during a campaign of civil disobedience to defy the ban and win back the right to protest outside the Embassy.

Of course, much might depend on how the Metropolitan Police Service interpreted arrests being 'in the vicinity of the South African Embassy'. Certainly, we would expect the figures quoted by City Group to be higher, precisely because they would include *all* arrests incurred during the Group's campaigning (including those on demonstrations elsewhere) and including those where charges were dropped.

The list of charges presented here is also of interest. Although many of those charges are fairly common in the context of policing protests and demonstrations, the charge relating to 'street trading' is a little less orthodox and relates to the police's attempt to stop picketers from collecting monetary donations on the street from the public. As we charted in Chapter 5, the police often stretched the credibility of the law (being 'unruly' in their own ways) in order to attempt to restrict the operation of the Picket. This was most intense during the first year or so of the Picket, but never totally subsided.

In January 1990, just weeks before the Non-Stop Picket would end, an internal document was written for City Group's committee entitled *Record of Police Harassment on and around the NSP, 28th September 1989–4th January 1990*.[3] It was written by Mark Farmaner, the Picket Organiser at the time, and was a detailed record of the Metropolitan Police's treatment of anti-apartheid protesters during that four-month period.

No single incident in September 1989 inspired Mark Farmaner to start monitoring police activity on the Non-Stop Picket, but he recalls that harassment was increasing at the time, and one or two officers were perceived to be acting over-zealously in their policing of the Picket.[4] He thinks that the City Group legal team recommended compiling the report believing it might be useful in pending court cases. It seems likely that responsibility for policing the Non-Stop Picket shifted around that time, in preparation for the opening of the new Charing Cross Police Station. For most of the Picket's duration, the protest was policed by officers from Cannon Row Police Station (with their AD badge numbers); but all the officers referred to in the January 1990 report have CX numbers.

As we outlined in Chapter 4, on each shift on the picket, one protester was designated as the Chief Steward. The Chief Steward served as the main (and, officially, the only) line of communication between the Picket and the police officers protecting the South African Embassy. In this role, they recorded details of any incident on the picket in a notebook that was passed from one Chief Steward to the next. During the period recorded in Mark Farmaner's report, at the end of each shift, the Chief Steward completed a report about their shift, noting who had been there, any new members recruited, any shortages of publicity materials that needed replenishing, as well as any incidents involving the police during the shift (Davies 2012). These reports were collected by the Picket Organiser daily and taken to City Group's office for action. Mark Farmaner's report on police harassment was collated from the Stewards' notebook and the reports completed at the end of each shift. Interestingly, several of the incidents recorded in the report demonstrate a renewed attempt by the police to undermine the role of the Steward as the point of reference for the Picket.

Taken in isolation, many of the incidents it records are quite petty, but as a single document, it makes grim reading. In total, it records (and cross references) 45 police officers who were present at incidents involving 40 separate protesters. Most of the officers cited in the report were only present at a single incident; a handful had five or more incidents attached to their badge number.

The specific incidents recorded range from petty insults and name-calling to the use of arrest under the Prevention of Terrorism Act to politically harass Irish picketers (Hickman 1998; Hillyard 1993). Some of the name-calling was simply personal and a little absurd – threatening to arrest one picketer because "he had an ugly face," or another because he "had bad breath." Frequently, there were racist and homophobic overtones to the incidents. One young Asian protestor, Selman, was racially taunted on several occasions, and officers stood in front of him on the evening of 10 October 1989 talking loudly about "monkeys." This event and others on (and off) the Non-Stop Picket around this time undoubtedly impacted on Selman's perception of the police:

> I think when it comes to the police I would say unreservedly they were bastards. They were politically motivated. They were not of apiece but largely racist. They had a deep-seated hatred of us and what we were doing. There were individual officers who were utterly rogue, and they were by and large, those rogue officers were by and large not controlled by their senior officers. ... Now I have a personal insight into this as somebody who grew up in London as a young Asian boy and then became a lawyer afterwards, and I think this is a very much a pre-McPherson police force, a pre-McPherson Met, because that was very much of apiece with my interaction with the police outside of the Picket as well. They were institutionally racist and in a lot of cases they were personally racist as well.[5]

Ruby told a similar story of being confronted by the realities of police racism through her experiences on the Non-Stop Picket:

> I hadn't experienced racism really I don't think ... In fact it was through getting arrested on the picket, that was my first experience of it [racism], and that's why it was quite a shock, and also that I'd never been called names like I'd been called by the police when I was arrested. You know, I'd never been called 'Paki' and it was a policeman.[6]

There were many other, similar incidents. On 5 December 1989, an officer followed a young black picketer, Danny, up and down the pavement repeatedly asking "what's that smell?" and making ostentatious sniffing noises.

On 28 September 1989, an officer arriving at the start of his shift greeted the predominantly male group of picketers present with a call of "hello girls; how are you?" in an affected camp voice. At other times during this period, gay and bisexual men on the picket[7] were told they "had a nice profile," called "paedophiles," and blown kisses. Although picketers from a range of social backgrounds were harassed during this period, those who were most vulnerable to police attention, and who were targeted repeatedly, appear to have been black and Asian picketers, Irish protesters, and gay men. Many of those who appear to have been targeted were younger picketers who were in the process of taking on more responsibility on the picket and within City Group. This may not have been deliberate, but it seems unlikely that it was coincidental. These small acts of harassment, these micro-aggressions, seem to have been used either to provoke arrestable offences or more generally as acts of attrition intended to undermine the confidence of younger protesters taking on leadership roles and test their resilience.

Mark Farmaner's report did more than just catalogue overly zealous policing. Through the act of compiling the report, the mundane records of daily life on the picket during those months were transformed. The individual entries in the Stewards' logbook and shift reports, when sifted, selected and placed beside each other in a systematic manner were used for political effect (Davies 2012). Patterns emerge from the document showing how micro-aggressions were directed at individuals and groups of protesters; small acts of harassment were made intelligible as part of a systematic approach to policing long-term protests. In these ways, the report would have had effects within City Group. It reminded activists that the policing of the Picket was political and would have shaped how they continued to relate to the police. In hindsight, some former activists have reflected that the tense relationship between picketers and the police may (counter-intuitively, perhaps) have contributed to the longevity of the Non-Stop Picket. Dominic, for example, questioned: "I wonder how much, to some extent, we thrived on getting picked up by them all the time and winning?" (cf. Halvorsen 2017). Given that City Group would often help protesters to sue the Metropolitan Police for wrongful arrests, if they were acquitted or had been found not guilty, with the resulting compensation shared between the Group and the individual, in many ways, the over-zealous policing of the Non-Stop Picket directly helped to fund the Group's anti-apartheid campaigning in later years. The Group's politicisation of its relations with the police also helped to reinforce a culture of political organising that took security seriously and sought to undermine low-intensity intelligence gathering targeted against it (Kitson 1971).

Organising (through) legal support

By locating the Non-Stop Picket outside the South African Embassy, City Group was able to frame all attempts to curb its protests as an example of the British state acting in the interests of South Africa's apartheid regime. This confrontational stance helped persuade many activists of the validity of risking arrest in the pursuit of their anti-apartheid campaigning. Although this political justification for confrontational and disobedient practices was important, one of the main reasons that so many City Group activists were prepared to be arrested was the quality of legal support that the Group provided for its supporters. This, in turn, also contributed to the longevity of the Non-Stop Picket because picketers knew they would be supported when they stood their ground to defend the integrity of the Picket and its core practices.

City Group tried hard to ensure that all participants in the Non-Stop Picket were aware of their legal rights and knew how to behave and what to do if they were arrested. It was the responsibility of the Chief Steward on each picket shift to keep a contemporaneous record of events surrounding arrests. If a picketer was arrested, it was the Steward's responsibility to ensure that the Group's office (or an out-of-hours contact) was made aware of the arrest:

> On City Group literature, there were instructions to demonstrators as to what to do in the event of an arrest. "If you get sent to Cannon Row, Phone 837 6050" The City Group office would then mobilise solicitors, members would go to the police station, to wait for release, supply bail etc. Often City Group members would phone around other members who would then phone the police station to enquire [after the arrestee].[8]

Although being arrested could be a daunting experience for young protesters (especially the first time), the briefings, support, and aftercare provided to picketers helped minimize the fear and apprehension. Picketers were so rehearsed in remembering the City Group office phone number as a contact in case of arrest, that 30 years later, many of our interviewees could still recite it without hesitation. Picketers were similarly well-versed in knowing their rights while in custody – a level of training that was not lost on the police:

> In the custody suites they behaved scrupulously correctly, but insisted on every right to which they were entitled. At court they always pleaded 'not guilty' and took up whole days, for months at a time, contesting evidence at trial and then appealing the results.[9]

Being aware of their rights, knowing that evidence had already started to be gathered for their defence, and being confident that the Group would provide them with sympathetic legal support throughout the court process, if necessary, meant that many picketers stopped worrying about arrests:

> Personally I think that our set up was brilliant. I never felt worried about the idea of being arrested. So when I was arrested, I knew exactly what to do. It was drummed into you.[10]

> I didn't have any fears at all about the legal support. I always felt it was, it was part of something bigger anyway. This wasn't about me, standing on a pavement. It wasn't Lorna Reid being arrested; it was the Picket being challenged.[11]

> So it's strange because you think being arrested would be one of your big fears, but the main concern at that point was like oh no, we've got to make sure that the Picket's restarted.[12]

City Group approached the legal defence of its supporters politically. It was one of the Group's principles that they would offer unconditional support to all those arrested on their protests and demonstrations, no matter how well they were known, nor what their political affiliation was. Just as the Group saw the policing of the Picket as a political attempt to curb effective protest against apartheid, so they ensured that a strong, well-organised legal defence was available to all defendants. City Group had an elected Legal Officer on their committee who would collate evidence and liaise with defendants, witnesses, and sympathetic lawyers to maximize the effectiveness of the legal support for arrestees. Frequently, City Group activists won their court cases precisely because their defence was better organised and more consistent than the evidence given by police officers for the prosecution.

By pleading 'not guilty', presenting a legal defence of their actions, and appealing guilty verdicts (when necessary), the Group maximized the costs of policing their protests. They were prepared to use public donations collected on the Non-Stop Picket to fund their legal cases.[13] These costs were frequently recouped later, as City Group was consistent in suing the police for wrongful arrest in response to acquittals and not guilty verdicts.

In addition to providing robust legal support to protesters at each stage of the legal process, City Group was quite prepared to use the courts in a more proactive manner. Whenever the Metropolitan Police seemed to be on the offensive against the Non-Stop Picket, using 'new' (in fact, frequently, ancient) laws in an attempt to restrict the protest, City Group's leadership would consult with supportive lawyers to devise effective means of resisting the police. Often, as one of those solicitors implied, this meant allowing well-orchestrated arrests to take place in order to take a test case through the courts in order to challenge the validity of the arrest and use of the laws being applied:

> My recollection and belief is that they thought quite hard themselves about how best to arrange things and to challenge the police and to get arrested, because essentially often they were deliberately getting arrested in order to be able to challenge the basis on which the police were trying to restrict their activities. And then as I say once they've been arrested we'd work together on how we were going to deal with it.[14]

A series of events from September 1986, six months into the Non-Stop Picket, serve to illustrate how City Group combined political campaigning, media work, and legal process to support its members. On Tuesday 16 September, five City Group activists were arrested outside the South African Embassy. The incident was reported in *Non-Stop News* the following week:

> At about 10.30am the police swooped on Hans. When Chief Steward Simone asked why, she was arrested, the arresting officer held her against the Embassy gates and pressed her from behind in a sexually obscene way.[15]

The next day, two more women picketers were sexually assaulted and abused during the course of arrests.

> Police Sergeant A33 led a violent sexual assault on Amanda Collins who was thrown into a police van face downwards. PS A33 got on top of her. Several police officers kicked and punched her while PS A33 sat on her. She was grabbed between the legs. At Cannon Row [police station] Amanda was put in a cell with several male police officers and assaulted. Cat was also thrown into the van and abused. The police: 'You'd like a great big black one up you', 'No, she's a lesbian'.[16]

In total, 11 arrests were made on the picket over a two-day period. In addition to the violent sexual harassment and assault of these three women picketers, two black picketers were called "niggers" by the police officers while in custody, and a young gay man needed hospital treatment for the injuries he sustained during his arrest.

City Group responded to the assaults on Simone, Cat, and Amanda by calling a "Hands off Women Picketers"[17] protest outside Cannon Row Police Station on 24 September 1986, and mobilised the support of women's groups from around London to the protest. A writ was even issued against the Commissioner of the Metropolitan Police, and civil action was taken in the courts to challenge the treatment of women protesters.

This level of organisation, support, and solidarity was effective – City Group repeatedly claimed that fewer than 10 per cent of arrests on their protests eventually resulted in convictions. The robust nature of their legal support structures ensured that many picketers were prepared to risk (often repeated) arrests in pursuit of their anti-apartheid cause. In the remainder of this chapter, we examine some of the circumstances in which these wilful arrests took place (but we return to this theme in Chapters 7 and 9 to consider the longer-term consequences of arrests for those youthful activists). City Group may have stood against the interests of the British state, but it was very effective at using the courts as a political platform when necessary.

Direct action elsewhere

City Group's activism was not restricted to Trafalgar Square: Picketers took direct action against apartheid across England and toured the country mobilising solidarity (for those resisting apartheid in Southern Africa and for themselves). These extended campaigns of direct action away from the Non-Stop Picket included 'trolley protests' against the sale of South African goods in supermarkets across London, where activists filled trolleys with South African produce, took them to the checkout, and then refused to pay for them. At their most effective, these protests could tie up the majority of checkouts in a targeted supermarket simultaneously.[18] In a similar vein, City Group organised frequent occupations of the SAA offices in Oxford Circus through their 'No Rights? No Flights!' campaign.[19] Finally, City Group activists took direct action at sporting venues around the UK, including pitch invasions at various rugby and cricket grounds, in protest at sportsmen and women who had broken the sports boycott of South Africa (Maaba 2001). For example, on 30 January 1988, Andy Higginbottom and a small group of other activists ran onto the track in Gateshead to attempt to prevent the South African runner Zola Budd (who had rapidly acquired a British passport in order to circumvent the sports boycott) from completing her race.[20] This was not the first, nor the last, time City Group attempted to disrupt one of her races.

There was a particular logic to organising actions of this kind – although City Group called on the British government and the wider international community to impose tough sanctions against the apartheid regime in South Africa, they were committed to building grassroots boycotts of South Africa. Their boycott actions were intended to inspire a wider grassroots boycott movement. In doing so, supporters of the Non-Stop Picket heeded the call for people's sanctions, issued in the summer of 1986 by James Motlatsi and Cyril Ramaphosa, two leaders of the South African mineworkers' union, during a visit to Britain. City Group propaganda frequently cited a call to action that they ascribed to these two South African trade unionists: "If your government will not impose sanctions, then you much act over their heads."[21] Their supermarket actions in support of a consumer boycott were usually targeted in ethnically diverse, working-class communities in inner London, where the Group anticipated they might find an audience and support for their message.

For these actions to be successful, activists used their bodies in particular ways to occupy space or block certain activities from taking place. They relied on their bodies' potential for intense speed (or slowness) to get into place to protest, evading capture by police officers or security guards, or to hamper their removal from the SAA offices or the cricket crease (at Lords' and elsewhere). Frequently, they would disguise their bodies or mask their identities in order not to appear unruly or out of place, thereby enabling their planned unruliness. Very often, these disguises could take absurd forms – so much so that it is a wonder they were not spotted and caught out more frequently. Timing was everything – protesters waited patiently for the optimum moment to act and coordinated their arrival at an action carefully (as Norma Kitson taught them, drawing on her covert work in South Africa

136 *Being unruly*

in the 1960s, to arrive too early could draw attention to oneself, but being late could leave others exposed and vulnerable). As should be clear, to create the potential to use their bodies in unruly ways, picketers exercised a high degree of calculation and control over their bodies.

The theatricality of these actions could be very effective in how they challenged bystanders to think about apartheid. It was not uncommon for supermarket trolley-push actions to inspire and empower other shoppers

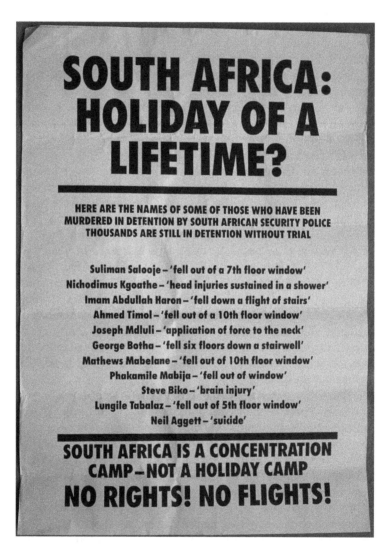

Figure 6.1 City Group poster produced for the *No Rights? No Flights!* campaign, 1988. Source: City Group.

to challenge the store managers about their continued sale of South African produce. Bystanders could be provoked to review and reconsider the goods they had in their shopping baskets and to buy non-South African alternatives. Of course, these protests never attracted universal support, and other shoppers would also erupt in frustration and anger because their Saturday morning shopping trip had been disrupted in this way. If the Non-Stop Picket was an extraordinary protest directed at the representatives of apartheid in Britain, these actions for people's sanctions took resistance to apartheid directly into the everyday lives of ordinary Londoners.

There were two campaigns of direct action that took place off the Non-Stop Picket, which have stuck in the memories of former anti-apartheid campaigners. They were the 'No Rights? No Flights!' campaign against SAA in 1988 and the protests against Mike Gatting's 'rebel' cricketers the following year. These actions are perhaps most memorable due to their spectacularity and the numbers of people who participated (directly or indirectly) in these two extended campaigns. We examine these two campaigns in more detail here to explore how City Group activists deployed playful forms of protest to disrupt the "British collaboration with apartheid" in its many distributed forms (Massey 2008) and draw attention to their cause (Figure 6.1).

No Rights? No Flights!

On 24 February 1988, in response to increasing protests inside the country, the South African government banned 18 anti-apartheid organisations, including the United Democratic Front, the Azanian Peoples Organisation (AZAPO), and the South African Youth Congress, as well as many local groups. The Congress of South African Trade Unions (COSATU) was banned from all political activity and restricted to trade union work alone (Lester 1998). In response, City Group escalated its existing 'Participate in '88' programme of action by launching a new campaign entitled 'No Rights? No Flights!', which targeted tourism to South Africa and particularly the state-owned SAA.[22] City Group had occasionally occupied the SAA offices since the early 1980s, but this was their most sustained campaign against them.

On one occasion, on 4 March 1988, City Group members occupied the offices of SAA in Oxford Circus. Standing in the window of the airline's offices, the occupiers held hand-drawn posters naming the recently banned anti-apartheid organisations in South Africa. A report in the March 1988 issue of *Non-Stop Against Apartheid* described the occupation in the following terms:

> Our placards called for the people's organisations to be unbanned, and our South African freedom songs rang in the ears of the international operators on the line to Johannesburg and Pretoria. People passing SAA raised their fists in solidarity and took our leaflets.[23]

The campaign continued throughout the first half of 1988. In early June 1988, six activists were arrested for 'highway obstruction' whilst continuing their 40-minute protest *inside* the SAA office. Presumably, the police argued that the visibility of the protest through the plate-glass windows of the airways' offices had attracted a crowd of bystanders that was itself obstructing the highway. Just as the Non-Stop Picket was highly visible in Trafalgar Square, so the position of the SAA offices on Oxford Circus allowed the Group to bring militant anti-apartheid activism to one of the busiest shopping streets in Britain.

Throughout the campaign, the SAA offices were frequently closed several times in a single day through multiple occupations. As the SAA security staff began to recognise repeat 'offenders', activists needed to utilize more and more imaginative disguises to enable their initial access to the premises. During one protest (to mark South African Women's Day in 1988), a large party of women, varying in age from their mid-teens to their seventies, successfully entered SAA dressed as nuns and a class of convent girls.[24] At other times, women of disparate ages entered in 'mother and daughter' combinations. Other disguises were also used, and some protesters would significantly alter the cut and colour of their hair, as well as adapting their apparel, in order to mask their identity. Many of the activists we interviewed took delight in remembering how they had donned smart suits to present themselves as businessmen and women; or, in one case, as a priest.[25] They variously remembered these protests as "slightly silly" because "we only had to look remotely respectable and we got in." For Helen Marsden, the protests were "very creative" and "like doing fancy dress."[26]

On 2 July 1988, City Group had their most successful day of action against SAA. Three separate groups of protesters managed to occupy the public offices of the state-run airlines calling for a full boycott of apartheid South Africa. Two further groups were lined up to take the same action. In the face of these successive waves of protest, the airways' offices were eventually forced to close for the rest of the day.

The action was highly organised. Each of the five groups of (potential) occupiers was led by one experienced volunteer who acted as their 'Chief Steward' (in a similar vein to shifts on the Non-Stop Picket). They had the role of taking tactical decisions and liaising with the police. Within each group, one person held a camera to record the action. Supportive photo-journalists accompanied two of the groups. Finally, legal observers and other supporters shadowed each occupation group to witness their actions and distribute leaflets about the protest to passing members of the public. Through these combined roles, at least 30 different activists took part in the actions that day.

Kathy was amongst the first group to enter the SAA offices that morning. Her witness statement records the events in the following terms:

At approximately 10am on Saturday 2 July 1988 I and three others entered South African Airways in Oxford Circus. [...] Immediately we were admitted our protest began. All four of us took out posters of "Sharpeville 6 Must Go Free," held these up in the windows and we began chanting anti-apartheid slogans.

When our demonstration commenced the SAA Security guard ushered all the staff at the front desk through a door at the rear of the premises. The Security guard remained in the front office with us, but did not approach us or attempt to speak to us at any time.

After several minutes five or six police officers arrived. An Inspector asked us to leave. We all refused. Steve stated to the Inspector that this was a peaceful and lawful demonstration and asked if we were being arrested. The Inspector informed us that the police had been called in by SAA personnel to assist with our ejection from the premises. Steve [Konrad] enquired of the Inspector if we were being arrested and if so on what charge. The Inspector said that if we did not leave we would be charged with police obstruction. We all sat down on the floor and refused to leave.

Alwyn was first to be physically dragged out of the premises. Steve was then taken out of the double doors of the SAA offices. Adenike was removed thirdly from the premises. I was the last to be physically removed. I assumed at this point that I was being arrested as I had refused to leave of my own accord. However, when I was placed on the pavement outside, I found that only Steve had been arrested. Adenike, Alwyn and myself continued our demonstration outside SAA.[27]

The second occupation group consisted of two City Group activists, Cat and Simon, and two supportive local councillors from the Labour group on Camden Council (one of whom used a heavy electric wheelchair). In her witness statement, Councillor A records the events of the second occupation:

At about 10.50am [Councillor B] and I approached the SAA door. The security guard asked why we wanted to go in. I said I needed to obtain details of access for people with disabilities in South Africa.

He then let us in and we looked at some leaflets. Cat and Simon followed us in a minute after but were trapped in the lobby between both doors, then started singing and we joined in and we held leaflets up to the windows. We did this for about 15 minutes. The guard asked us to leave but we declined.[28]

Councillor B's witness statement explains what happened next:

The police then arrived. The policeman said that the guard had the right to evict us and could use police assistance if necessary. We still said no.

> I sat on the floor and held hands with [Councillor A]. The policeman and guard then lifted me up while I went limp. I was taken outside the building, followed a few minutes later by [Councillor A]. We then gave out leaflets, [Councillor A] then placed herself in front of the doors and was eventually arrested.[29]

It took the police some considerable effort to move Councillor A following her arrest, given the bulk and weight of her wheelchair once the brakes were applied. Cat was also arrested following her eviction from SAA but Councillor B was not. Simon and three members of the support team were all arrested outside the building. By this time, the volunteers in the City Group office were working hard, calling the media about the arrests and calling high-profile supporters of the Group, such as Tony Benn, requesting that they phone West End Central Police Station to check on the well-being of the arrestees.[30]

The third occupation group, who were due to enter the SAA offices at midday, was not able gain access to the premises. It seems likely that their planned action had to be aborted because the police were still present outside the SAA offices following the previous occupation. In the original plan for the day, there were to have been five occupations in total, with teams due to enter the building at 1 p.m. and 2 p.m. too. In the end, the remaining two groups of volunteers were reconfigured, and a final group entered the building in the early afternoon. Of the five members of that team, three made it inside the SAA offices. All were ejected, but only Liam was arrested.

By this point in the day, City Group's normally tight and efficient legal support processes appear to have unravelled slightly. The log of events recorded by the volunteers in the office suggests, with some frustration, that the police at West End Central were keeping them informed of arrests (and releases from custody) faster than the legal team on the ground could manage. In total that day, eight anti-apartheid activists were arrested in the course of three occupations of the SAA offices.

These events demonstrate the commitment of City Group supporters to being 'non-stop against apartheid'. It took a high degree of coordination to plan a day of action in which up to five groups of activists would have occupied the offices of SAA. That up to 20 volunteers were prepared to risk arrest that day demonstrates the centrality of direct action to City Group's approach to anti-apartheid solidarity. The support structures that City Group put in place, in terms of legal observers, logistical support from the office, and a team of people willing to keep phoning the police station to check on the welfare of those arrested and wait outside to greet them on their release, added to the confidence of those risking arrest. City Group offered practical solidarity to its own supporters, just as it took action in solidarity with those resisting apartheid in South Africa.

In addition to these overt actions to temporarily close the SAA offices, the covert group 'Red Reprisal' claimed responsibility for superglueing shut the doors to the Airways' showroom on one occasion and smashing a large plate glass window in the building on another.[31]

Stop batting for apartheid

During the summer of 1989, a cluster of City Group activists attended several cricket matches around London and the Southeast. They were not there to enjoy the game; they were there to protest against Mike Gatting's involvement in a rebel cricket tour of South Africa in contravention of the sports boycott.[32]

In organising these protests, City Group took their lead from the South African Council on Sport (SACOS) that there could be "no normal sport in an abnormal society" (Nauright 1997). They also drew on the legacy of the Stop the Seventy Tour campaign against South African cricket and rugby tours of Britain (Fieldhouse 2005); and the Halt All Racist Tours (HART) campaign in Aotearoa/New Zealand (Pollock 2004; MacLean 2010), which were both prepared to take mass direct action to enforce the sports boycott of South Africa. Accordingly, City Group activists set out to disrupt those cricket matches where Gatting and his fellow 'rebels' were playing. While some activists stayed outside the cricket grounds protesting, distributing leaflets, and trying to convince the cricket fans about the validity of the sports boycott of South Africa, others would buy tickets, enter the ground, and then, when the moment was right, seek to disrupt the rebel players' game.

On 13 August 1989, 14 activists invaded the pitch at Lords' Cricket Ground to stop Mike Gatting from playing. In total, eight of them were arrested for this action, with the rest being escorted from the ground. As with the SAA occupations, the activists donned disguises, dressing up in an attempt to fit in with the crowd. Georgina remembers that she was wearing "this girlie blue, properly blue dress and some little tiny shoes [with] my hair up in a ponytail." Her former school friend, Sharon, remembered their pitch invasion "with great fondness," recalling how they and others modified their appearance and behaviour in order to execute their action:

> They were expecting demonstrators so we had to kind of dress up as cricket supporters. Myself and Georgina were wearing frocks and looking like ladies, and Dom and Daniel were wearing jackets and ties. We had to sit apart in groups so that we could stagger the pitch invasion. I remember Dom pointing with his umbrella to another protester outside Lords and muttering "riff-raff" under his breath. When Mike Gatting came out to bat, we had to jump over the barrier and run to the crease. We actually got to the crease and I think either Dom or Daniel pulled out the stumps.[33]

Georgina was surprised that they managed to reach the centre of the pitch to protest.

> We thought we'd be stopped. We actually got to the crease. We were like oh my god this is ... Oh you're just waiting for that moment and it's like your heart starts to go and you're just ready to jump up because that once you're over that's it, you've got to run as far and as fast as you can and make sure that you've got the banner out, you know, so that everyone can read it and see it.[34]

The first quartet of protesters to invade the pitch was escorted from the ground. But further groups continued to replicate their protest and disrupt the match whenever the 'rebel cricketers' attempted to play. As with the first group, the later groups ran onto the pitch in pairs. Like Sharon and Georgina, they tried to reach the crease together, but sometimes that was not possible – leading, in one case, to a blind protester running across the pitch unescorted:

> We sat for hours and hours and hours on the cricket and then ran onto the pitch, and I ran on with Solomon and he was blind, and I said can I let go of you, and he was terribly brave, he said yes and he kept running. I thought he'd stop running, but he kept running and I got to the stumps and pulled them up.[35]

In the process, Solomon dropped his 'white stick'. While Richard (and the others who went onto the pitch with them) was arrested, Solomon was left outside the cricket ground's gates by the stewards. Without his mobility aid to assist him, he still managed to find his way from the sports ground to the police station where his comrades were detained to wait in solidarity for their release. Solomon's commitment to keep running and his resourcefulness in negotiating a route to the police station draws attention to the embodied and emotional intensities of engaging in direct action like this. His case might be extra-ordinary, in some ways, but it has parallels in the many ways that the protesters took pleasure in disguising their intentions, and displayed determination in motoring to the crease and evading capture on the way. Of course, as Richard and Georgina imply, the experience of thrill and excitement was also accompanied by anxious anticipation and extended periods of boredom inside the ground.

Two weeks after the action at Lords, on Sunday 27 August, seven supporters of the Non-Stop Picket were arrested during a similar action at the Kent County Cricket Ground in Folkestone. One of them, Dominic, was assaulted by an outraged cricket fan as he was escorted through the members' pavilion under arrest. The protesters were eventually released without charge, but Dominic's assailant was charged and taken to court. By this point in the campaign, the members of the sanctions-busting cricket tour were beginning to become familiar with their opponents. On this occasion, their primary target was the Kent player Chris Cowdrey. When he came into bat, the protesters also took to the field:

But me and Helen got to him and Cowdrey was like oh you again and was sort of slightly cheery, you know, we had a little bit of a barney, and then we got arrested.[36]

Once they were in detention, so many of the protesters' supporters rang the local police station to enquire about their well-being that the exasperated custody sergeant started to give the automatic response that "the people from the cricket protest will be released as soon as the match is over" as soon as any caller was put through to him. This volume of solidarity may have contributed to the release without charge of all the Folkestone protesters:

In addition to their protests and direct actions, the non-stop picketers once again utilized the court system to amplify their cause. As the court case for the Lords' arrestees drew near, City Group's legal team decided to call Mike Gatting as a witness and issued a witness summons to compel his attendance in court. Danny Simpson, one of the Group's solicitors in this case, explained how this arose.

> Their trial, by coincidence, was listed for about day two of the actual tour. I spoke to Gatting and Ramprakash, who were both on the pitch [at the time], and neither felt threatened. Gatting said he was holding a big lump of wood in his hand and was pretty nifty with it, so he didn't fancy anyone's chances who came anywhere near him. I was able to get a witness summons for him, which we then served, briefly threatening the tour. Unfortunately, his legal team went to court and got it quashed – something I regret because it was a genuine summons for a key witness whose evidence was relevant to the question whether the behaviour was threatening.[37]

This final twist in sports boycott campaign re-emphasizes the point we have made throughout this chapter – while City Group was committed to using direct action and civil disobedience in pursuit of its anti-apartheid cause, these tactics were frequently combined with more traditional campaigning practices and a bold approach to the legal system.

Empowerment through unruliness

Throughout this (and the previous) chapter, we have outlined the distinctive practices and culture of protest, defiance, and solidarity created on and by the Non-Stop Picket of the South African Embassy in the late 1980s. The Non-Stop Picket survived for nearly four years because its regular participants (and its more peripheral supporters, when necessary) were prepared to stand their ground and risk arrest in order to establish and maintain a right to protest on their own terms, in the way they saw fit.

In these ways, individuals not only learned how to use their own bodies in unruly ways but also, through repeated joint actions with other picketers, developed tacit knowledge of how other bodies would respond in particular circumstances. This shared experiential knowledge could assist the Group to pull off spectacular acts of civil disobedience, when the combined bodies of protesters acted in concert to out manoeuvre the authorities. One such occasion was 6 September 1989 – the day of what turned out to be the last racially segregated general election conducted under apartheid in South Africa – which was a peak experience for many picketers.[38] On that Wednesday evening, up to 1,000 people swelled the Non-Stop Picket in an angry challenge to apartheid. That large crowd ended up blocking the traffic in Trafalgar Square for more than two hours during the evening rush hour. To block the road, groups of picketers used their bodies in unruly ways – darting through lines of police to enter the road, risking their safety by lying in front of cars and buses, linking arms, and going 'limp' to hamper police efforts to remove them from the roadway (and then jumping straight back in again). Despite the significant disruption caused that evening, the police were very restrained on this occasion – refusing to give City Group the publicity that hundreds of arrests outside the Embassy would have garnered. Once the crowd had secured the road, a band set up on the tarmac and an impromptu street party ensued for the next hour and a half. Eventually, the protesters vacated the road en masse, at the time of their choosing, singing *Nkosi Sikelel' iAfrika*.[39] The police's response to this event highlights the relational nature of policing and protest on the picket – picketers would try out new protest practices, using their bodies in new ways to confound established police tactics, but they could never fully anticipate how the police would respond to their unruliness. At times, a restrained response from the police could confound the picketers' political objectives. It attracted some media attention, but not the volume of coverage, nor the political capital, that mass arrests might have done.

A study of the Non-Stop Picket's practices of direct action against the representatives of the apartheid regime has been central to this chapter. We have argued that through this non-violent, but confrontational, political stance, the picketers learned to think and act against the (British) state, using their bodies in (sometimes mundane, but often extra-ordinarily) unruly ways. Picketers used their bodies on and off the picket to physically block the normal functioning of apartheid's representatives in Britain, trade with South Africa, and those who sought to break the international sanctions against apartheid. The space occupied by the Picket, on the pavement in Trafalgar Square, and the vibrant noisy culture of the protest allowed teenage protesters and others the opportunity and freedom to express themselves in ways that might have been constrained in other areas of their lives. The entangled practices of solidarity activism and 'growing up' are the focus of the next chapter.

Notes

1 Email correspondence with Sally O'Donnell.
2 Metropolitan Police Service Freedom of Information Request Reference No: 2013030000771, released 10 September 2013.

3 M. Farmaner (1990), *Record of Police Harassment on and around the NSP, 28th September 1989–4th January 1990*. Internal report for the City of London Anti-Apartheid Group Committee.
4 Interview with Mark Farmaner.
5 Interview with Selman Ansari.
6 Interview with Ruby Noorani.
7 Gavin was one of those mentioned in the report as having been subjected to such low-grade abuse and harassment.
8 Interview with Sharon Chisholm.
9 Email correspondence with David Gilbertson.
10 Interview with Georgina Lansbury.
11 Interview with Lorna Reid.
12 Interview with Grace Livingstone.
13 Some picketers and other supporters of the Non-Stop Picket expressed discomfort at the time that this was seldom made transparent to those giving the donations. Although legal fees were not directly addressed, the City Group papers contain a letter (dated 29 September 1987) from Ravinder Bhogal, the black workers' group coordinator at Brent Council, asking for a City Group representative to attend a meeting to explain how their donations for black people in South Africa were being used.
14 Interview with Danny Simpson, 26 February 2014.
15 'Racist, sexist thugs', *Non-Stop News*, No. 11 (19 September 1986), p. 1.
16 Ibid.
17 'Metropolitan Police sexually harass women picketers', *Non-Stop News*, No. 12 (10 October 1986), p. 1.
18 In October 1986, City Group coordinated simultaneous trolley protests at seven supermarkets across ethnically diverse areas of inner London. This action was promoted in advance, in an unsigned article called 'Peoples Sanctions Now', *Non-Stop News*, 12 (10 October 1986), p. 2.
19 City Group (1988), *No Rights? No Flights!* A5 leaflet (July); City Group (1988), 'No rights – no flights. City Group closes SA Airways', *Non-Stop Against Apartheid*, 27 (April), p. 3.
20 'Zola race protest', *Sunday Sun*, 31 January 1988.
21 'Peoples Sanctions Now', *Non-Stop News*, 12 (10 October 1986), p. 2.
22 City Group (1988), 'Participate in 88! Free the prisoners now!', *Non-Stop Against Apartheid*, 25 (January/February), p. 6. City Group (1988), 'Participate in '88. Join the action. No rights – no flights', *Non-Stop Against Apartheid*, 26 (March), p. 1.
23 'No rights – no flights. City Group closes SA Airways', *Non-Stop Against Apartheid*, 27 (April), p. 3.
24 Sigga (1988), 'All women's occupation', *Picketer's News*, (12 August), p. 2.
25 Interview with Mike Burgess.
26 Interview with Helen Marsden.
27 Witness statement of Kathleen Fernand in relation to arrest at South African Airways, 2 July 1988.
28 Witness statement of Councillor A in relation to arrest at South African Airways, 2 July 1988.
29 Witness statement of Councillor B in relation to arrest at South African Airways, 2 July 1988. R. Roques (1988), 'No Rights! No Flights!', *Non-Stop Against Apartheid*, 28 (June), p. 2.
30 Alongside the witness statements relating to these arrests in the City Group archive is a typed document "Office diary 2 July" logging all of the phone calls in and out of the group's office on that day.

31 Mackerel (1988), 'Paint Job', *Picketer's News*, 2 (27 May), p. 1; City Group (1988), 'Blood on his hands', *Picketers News*, 8 July, p. 1; City Group (1989), 'Chuck Paint and Stripper Shock!', *Picketers News*, 2, 1 (21–26 January), p. 1.
32 City Group (1989), 'No balls!', *Non-Stop Against Apartheid*, 35 (July/August), p. 7; D. Thackray (1989), 'Stop the Tour', *Non-Stop Against Apartheid*, 36 (October/November), pp. 6–7.
33 Interview with Sharon Chisholm.
34 Interview with Georgina Lansbury.
35 Interview with Richard Roques.
36 Interview with Dominic Thackray.
37 Interview with Danny Simpson.
38 Interview with Jacky Sutton; interview with Rebecca Copas; interview with Claus; interview with Francis Squire; interview with Simon Murray; interview with Deirdre Healy; interview with Nick Manley; interview with Andre Schott; interview with Mark Farmaner.
39 Composed as a hymn in the 1890s by Enoch Sontonga, a Methodist preacher, the song – its title meaning 'God bless Africa' in Xhosa – was the official anthem of the ANC during the apartheid era and a powerful anti-apartheid symbol. Part of the melody and lyrics has subsequently been incorporated into the South African national anthem.

7   Growing up through protest

Young people and even children were central to sustaining the Non-Stop Picket (Brown and Yaffe 2016). While students and the unemployed (sometimes homeless) youth participated there day and night, on weekends, and during school holidays, younger teenagers and children played a prominent role. Although they learned much from older activists, this intergenerational exchange worked in both directions (albeit, not always without conflict and misunderstandings). The Non-Stop Picket was a space where young activists were encouraged to take political responsibilities in various ways, and given the opportunity to develop new skills. Their youth gave them the energy, certainty, and determination to maintain a non-stop protest over a four-year period. Former picketers also remember the fun, excitement, and strength that came from taking collective action about something that mattered (to them and to the world).

Through their shared commitment to anti-apartheid solidarity, these diverse young people grew up together and learned to cope with the everyday pressures of youth. The anti-apartheid movement was not a backdrop to these young people's lives; they grew up *through* their political engagement. The act of standing together outside the Embassy regularly, often for hours at a time over an extended period, fostered strong social solidarity amongst the Group. This had many implications for picketers' lives, and close familiarity also generated levels of trust between them that enabled them to be more effective when taking action together. The comradeship that developed amongst them enabled them to help each other overcome social isolation, depression, unrequited love, conflicts with their parents, and trouble at work or school. Although their activism entailed extending solidarity with distant others, the everyday life of the Picket fostered more immediate experiences of collective support and enabled participants to see beyond their own positionality and (often limited) prior life experiences. The Picket provided a safe and supportive milieu in which to experiment with different identities and ways of being.

This chapter argues that young activists' political commitments are always entangled with the everyday politics of youth; that (in the context of the Non-Stop Picket) to practise solidarity was also to develop competences and resources that contribute to the practice of growing up. Although this chapter focusses on the experiences of teenagers and young adults, it also argues that 'youthfulness' and practices of 'growing up' are relational and not age specific. Several picketers who joined the protest in their late twenties and thirties described how their involvement with the social and political life of the Non-Stop Picket gave them opportunities to 'really grow up'.

Young picketers

It is perhaps tempting to assume that most young people who engage in activism around major geopolitical issues (in a country like Britain) are relatively privileged and highly educated. That description fits many of the young people who participated in the Non-Stop Picket, but not all of them. While a significant proportion came from professional middle-class families, many grew up in more 'intermediate', aspirational, working-class families (the sons and daughters of skilled manual workers, technicians, and self-employed tradespeople). But there were also working-class black youth from inner city London, young migrant workers from across the globe, and those who found themselves living rough on the streets of Central London for various reasons.

The 1980s were a mixed time for young people. By 1983, unemployment nationally topped 3 million (but was probably higher) (McSmith 2011). Even in relatively prosperous London, unemployment disproportionately affected young people (and particularly black youth) (Abercrombie et al. 1988). Although the experience of unemployment could be depressing, it did create relative autonomy for those who chose to commit to activist (or artistic) pursuits (as they could afford to live in London and do these things whilst claiming benefits). But most of all, the 1980s was a decade of very rapid social, economic, and political change. Although Thatcher's supporters frequently pursued socially conservative policies, social attitudes as well as economic relations became more (neo)liberal during the 1980s. The young people who joined the Non-Stop Picket may have understood themselves as being in opposition to Thatcherism, but they could not escape the influence of the social changes occurring around them.

Georgina had grown up in suburban Southeast London. She joined the Non-Stop Picket soon after it started with her best friend Sharon. They were 16 and had both just left school. Georgina described herself at the time, in the following words:

> I was on the picket because I was already politically aware. I had a lot of half-formed ideas. I read an awful lot. I was aware of racism through the years ... I was definitely looking for something in that kind of direction. Both Sharon and myself were talking about it, we were very aware of it, we were looking, we had these discussions, people we knew were politically minded already and we wanted to do something and I think that that's part and parcel of the sheer joy of being young and looking around and thinking you've got this time as well.[1]

Growing up through protest 149

Like Sharon and Georgina, Andre was 16 when he joined the Picket in the summer of 1986. He was Dutch and was visiting relatives in London, with a friend from Holland, for the school holidays:

> We used to go out clubbing and stuff in Leicester Square, and then one night waiting for the bus, we saw some people standing there so we kind of went over, and this was in July 1986, and that was it. There were some nice people there, there were some young people and we just kind of had a friendly chat and basically they recruited me, you could say, for the rest of the holiday.[2]

City Group had been organising regular protests outside the South African Embassy since 1982. So, when the Non-Stop Picket was launched in April 1986, there were also a number of young people involved in the Group who had joined as a result of its previous activities. James had joined the Group a year or two earlier, as a result of a chance encounter with a key activist from City Group following his arrest on a demonstration in support of the Miners' Strike. He had grown up in a liberal Jewish family on the suburban fringes of North London and had been active in the youth section of his local Labour Party branch before getting involved in anti-apartheid activism. James played a key role in helping to launch the Non-Stop Picket, by which time he was in his late teens:

> I was entranced, hypnotised towards this incredible group of people who seemed so committed, so well organised, so knowledgeable, so passionate about what they were doing.... The immediate build-up was really exciting. I remember particularly planning meetings, because I was on the committee of the Group at that stage and, I'm not sure if I was a Legal Officer, I think I was a Legal Officer or at one stage maybe an officer without portfolio actually. Such a young age was quite humorous and even being a Legal Officer was quite humorous. I'm not sure I realised ... the complexities I was immersing myself in potentially.[3]

As the Non-Stop Picket continued over the next four years, several of these early recruits became skilled and respected organisers, and the Picket benefited from the influx of new 'generations' of younger teenagers. Ben joined the Picket in its final months (but continued to be involved in the Group's campaigning until the end of apartheid). He was just 13 when he started attending the Picket, following the example of his older sister. They had grown up in a middle-class family in North London:

> [Our parents were] gently political but Left leaning and definitely questions of race and equality I suppose were an issue within the family ... So I'd left primary school, I had started secondary school, which I really didn't like when I first arrived, and I was a chorister ... I was quite politicised and certainly at school I was quite proud of it, like the political

one about everything and sort of telling people off for being misogynist or racist or homophobic or whatever in an all-boys private school, and being a bit sort of annoying.[4]

There is a sense throughout the interviews quoted here of a group of young people who were brought up in families that took an interest in politics. As Gary recalled:

> I was brought up by, you know, Jamaican immigrants. I suppose not lower middle class, but like upper working class. My dad was an engineer before he dropped it. My mum was just like a housewife ... Not a normal kind of black, what people typically identify as a kind of black cultural family. My dad was very Left-wing. We had pictures of Stalin instead of Jesus in the house.[5]

The Non-Stop Picket was enticing for these young people and gave them the opportunity to become politically engaged on their own terms.

There is a danger of presenting these young people as 'extra-ordinary' – and, in some ways, of course, through their actions and commitments, they became so. But most were fairly ordinary children and teenagers who were juggling their political activity with all the pressures and concerns of the transition to adulthood. As Nicole, a sixth-form student, said:

> I was never able to commit as much time as I would have liked. I was doing my A levels initially, and being a tad boring, prioritised that I guess.[6]

Even so, as many former picketers reflected, whether they were in education, work, or unemployed, most of the young picketers did not have major life commitments. Their lives were not fully embroiled with the responsibilities and obligations of adult life, and it was easier to fit their other commitments around time on the picket. Because it was guaranteed that someone would be there, many picketers also turned up whenever they were bored or at a loose end – relishing having somewhere to hangout, with a purpose, with like-minded peers. For many activists, their peer group soon ceased to be defined by age, and close intergenerational friendships developed through the common experience of anti-apartheid solidarity activism.

If they had one thing in common, it was often a sense of being a bit lost, not fitting in in their everyday lives, and being restless with their place in the world (see Chapter 3). These were not necessarily traits unique to the younger picketers (several of the people who joined the Picket in their thirties, often after a divorce or similar major life transition, also expressed similar feelings), but we believe those feelings are often quite common for young people making the transition to adulthood. Rikke's experience typifies this – she is Danish, had come to London as an au pair, but joined the Picket just after she had walked out of that job:

I guess it was an outlet for some of my young and restless energy and the place I felt most at home at the time.⁷

It is arguable that, by leaving her home country to work abroad, Rikke had already expressed a youthful desire for independence and adventure. She soon realized that au pair work could not contain her 'restless energy', but she found her needs and desires to be better met in the broader context of collective political activism. In this way, the everyday politics of youthfulness and growing up were entangled with the young picketers' commitment to challenging apartheid in South African and Britain's economic and political support for the apartheid regime.

The Non-Stop Picket did, however, create an atmosphere where young people's political opinions and motivations were taken seriously. The protest was open, welcoming, and inclusive – it encouraged active participation. Keeping the Picket going, as a continuous protest, provided many opportunities for young people (including those barely into their teens) to take on various responsibilities. New picketers (of all ages) were encouraged to engage the public in discussion about apartheid. As they became regular, trusted participants in the protest, they might be encouraged to make speeches or lead singing and chanting on the megaphone; to contribute to organising a special protest; or to take responsibility for providing political leadership of a shift on the picket's weekly rota. In taking on these responsibilities (and being trusted by older people to do them), young picketers developed new skills, but also learned more about themselves.

Ben reflected:

> What was so interesting actually was the amount of autonomy and respect that even as teenagers you got from other people being there, there was something quite amazing about ... At its height I was doing nine-hour shifts on Saturdays, and that was partly because of the place and people were taking me seriously. And I think that may well have had something to do with what I was trying to become, you know, as much as the politics, although the politics was very important.⁸

He continued:

> I think about it now, as hugely important for me developmentally, and whether it was just the politics but also the emotional [support] and sort of being treated as an adult.⁹

In addition to the Non-Stop Picket, City Group activists organised protests and direct actions all over London (and beyond), attempting to disrupting Britain's economic, political, and cultural links with apartheid South Africa. For Ben, participating in protests in support of the boycott of South African goods was empowering and enjoyable:

> There were some really fun things. There's a shopping centre behind Russell Square, Brunswick ... And shoving don't buy apartheid stickers on fruit and veg, you know, all of that. I think it's brilliant for a 14-year-old, sort of *sanctioned mischief*.[10]

For young people, being on the picket was both serious and fun. They were taking action for a political cause that they believed was important. In the process, they were taken seriously by adults and given opportunities to develop new skills and confidence. Even so, a minority of those we interviewed remembered that this combined enthusiasm for fun and politics could sometimes lead to conflict with those adults who performed 'serious politics' in less frivolous ways. Mark Farmaner put it like this:

> So this big class difference came out onto the picket as well. Especially because I think some of them, because they liked to smoke joints, they liked to get drunk, they liked to do what young people do, and the RCG people in particular didn't like that. They felt they weren't serious and could be very patronising and condescending to them. They wanted disciplined people to behave in a certain way and stuff like that.[11]

Despite the occasional tensions over how political activists should embody their political discipline, for young picketers, their youth gave them energy, certainty, and determination to maintain a non-stop protest over a four-year period. That youthful lack of doubt was sometimes expressed with an earnestness that some of our interviewees cringed at as they reflected on it from the vantage point of early middle-age. Andre encapsulated this complexity when he reflected:

> At the time we were very convinced of our righteousness and our – how shall I say it – our cause, and we thought we were pure and other organisations were sell-outs. When I reflect on it I think I'm proud of many things that we've done, and many things we did were right ... However, I haven't got a clear political position on everything, I just look back and I see contradictions.[12]

The young picketers identified with the militant 'young comrades' who were resisting apartheid and attempting to decolonize South African society (Bozzoli 2004; Seekings 1996). They challenged the geopolitical perspectives of many in the British government at the time who still saw white South Africa as a bastion against the strategic interests of the Soviet Union and its allies in Southern Africa (Byrd 1988; Onslow 2009; Reynolds 2015; Shubin 2008). In London, the picketers understood their protests as a means of disrupting the smooth operation of the South African Embassy and, by extension, the economic and political links between the British and South African governments. Through their contact and interactions with members of the South African liberation movements, particularly those influenced by pan-African and Black Consciousness ideologies, young picketers contributed to the development and dissemination of subaltern geopolitical knowledge (Sharp 2013) that imagined a different role and new alignments for a post-apartheid South Africa.

Former picketers also remember the fun that came from taking collective action about something that mattered with other young people like them. The Picket provided a space in the centre of London where young people could sing, shout, test the limits of the law with the police, and debate the type of world they wanted to grow up in. Many participants found that enticing and empowering (Brown 2013). The close familiarity that developed between regular picketers helped generate levels of trust between them that enabled them to be more effective when taking action together. However, young activists' political commitments were always entangled with the everyday politics of growing up. Together they produced collective knowledge about geopolitical issues, but they also helped each other to find their own place in a rapidly changing world.

Youthful agency

> I do think that City Group were good at giving young people (or any people) a role in politics and helping them, enabling them to take part and organise it themselves, whereas the Anti-Apartheid Movement was more of a sort of official, it was for sort of established political figures or big names, and they organised fantastically big marches.[13]

Young people involved with the Non-Stop Picket appreciated the space it provided for them to play an active role in political campaigning. The Non-Stop Picket, the City Group office, and the campaigns that the Group ran were spaces in which young people were encouraged to take the initiative. To support and facilitate this environment, City Group worked hard to ensure that its members and supporters were well informed about the functioning of apartheid, the history of resistance to it, and current events in South Africa as they were unfolding. Through its weekly meetings, regular 'pavement universities' on the Non-Stop Picket outside the South African Embassy, as well as informal conversations between members, the Group engaged in informal popular political education about apartheid (Mills and Kraftl 2014).

On Sunday 7 December 1986, City Group held a day school at the Polytechnic of Central London on Marylebone Road. As the leaflet advertising this day school shows, the topics covered a range of themes linking the struggle against apartheid in South Africa with anti-racist activism in Britain.[14] There were sessions on the experiences of political prisoners in South Africa, the history of the armed struggle against apartheid, the political economy of British investment in apartheid, and the role of the Mozambique as a post-colonial country on South Africa's frontline. Each of these educative workshops was led and presented by a leading member of City Group – David Kitson spoke on the armed struggle, and David Reed discussed economics, for example.

The role of this day school (and similar events) was not, of course, solely educational. The strap line for the event was "Are you against apartheid? Join the action now!" The day school was intended to enthuse and activate City Group's supporters, to mobilise new supporters into anti-apartheid activity, and to sustain the commitment of existing activists. This was achieved through a series of short 'motivational' speeches (by Carol Brickley and Steve and Norma Kitson) and performances by City Group Singers. Alongside the educational workshops at the event, there were skills-based sessions designed to teach South African liberation songs and educate picketers about their legal rights when dealing with the police. There were also more practical workshops (mostly led by younger activists) designed to bring together specific interest groups – youth and students; trade unionists; women, lesbians and gays; and black and anti-racist organisations – to develop plans for future campaigning.

Although events like the December 1986 day school were important in allowing anti-apartheid activists to educate themselves and their supporters about the realities of apartheid in South Africa, they were not always popular with some of the young people who sustained the Non-Stop Picket. For those activists who engaged with these events, the workshops and their affective motivational components served to reinvigorate their activities. For others, these day schools and meetings were perceived to 'get in the way' and distract from the action of *doing* anti-apartheid activism. To reach and educate them, City Group deployed other tactics – from informal discussions on the picket to the production of the internal weekly newsletter *Picketers' News*, which was instigated in May 1988.[15] There were, however, multiple ways in which young (or other new) activists could take on and develop responsibilities on and around the picket. Many of these (and the encouragement to get involved) were integrated into the very practices through which the Non-Stop Picket was maintained. In what follows, we examine some of the ways in which young people accepted (and occasionally resisted) different kinds of responsibility within the Group. We also examine some of the more exceptional ways in which young picketers (although, admittedly, mostly young adults rather than teenagers) took responsibility for representing the Group at high-level meetings of the South African liberation movements and other international bodies. One of the specific organisational methods for encouraging the development of young activists was through the Youth and Students sub-group of City Group's committee. We close this section with an overview of the sub-group's work before examining in further detail some of their more ambitious (and memorable) achievements.

Taking responsibility

While members of the Non-Stop Picket contested the legitimacy of apartheid South Africa's diplomatic mission, they developed relationships of solidarity with members of the South African and Namibian liberation movements. As a result of the mutual trust established with members of the Pan-Africanist Congress (PAC) and the Black Consciousness Movement (of Azania) through the Group's solidarity activism, a small number of young picketers (more often young adults than teenagers, but not exclusively so) had the opportunity to represent City Group at conferences and meetings of these liberation movements (McConnell et al. 2012) and contribute to their creation of subaltern geopolitical imaginaries (Sharp 2013). In 1988, two young activists from City Group, Charine and James, (both in their early twenties at the time) represented the Group at a conference organised by the PAC's Women's Section in Dar es Salaam. They presented an official message of solidarity from the Non-Stop Picket and engaged in extended discussions around the conference. For James, this experience was

> Incredible! Not just in Dar es Salaam, I was in Harare with them too and yeah, and they looked after me, they met, greeted, transported me, looked out for me beyond looking after me, and wherever I went there were people with side arms or other methods of defending themselves and ourselves. It was really intense. And I met a ton of people while I was travelling there and attending the conferences and being able to meet with people who'd come out of South Africa for this particular meeting and many of the people who were living in exile and see people organising away from their homes and in exile either with or without their families and friends and away from their communities and being surrounded by the history and the struggle ... Being able to have really long detailed discussions about political philosophical (and extremely mundane and whatever) things with people in the PAC offices in Harare and Tanzania ... It was a real privilege to be able to do that.[16]

In contrast, Charine (who was a more experienced activist, having first joined City Group as a teenager during the 86-day picket for David Kitson in 1982) was more balanced in her assessment of the trip.

> It was *really* good. It was quite moving, but I don't think (even at the time) that I thought it was as well organised as it could have been and I remember being quite aware of the organising falling short of what events organised by the ANC had been. So I did see why, when I was there I could see why it was that the ANC had got further with its campaigning than the PAC had, because they weren't so well organised ... As big a fan as I was of the PAC, I could see that they were a lot less well-organised and a lot less well-oiled as a campaigning machine that the ANC were.[17]

Being involved in these high-level international meetings with members of the South African liberation movements allowed young picketers to develop deeper understandings of the movements they acted in solidarity with. However, these encounters could be daunting. Andre was another young activist who travelled internationally to represent City Group at 'diplomatic' meetings (Craggs 2014a; Dittmer and McConnell 2015):

> I once went to Geneva, there was some United Nations Special Committee Against Apartheid (I think they were called) and the Pan-Africanist Congress had special status with the UN ... And, as such, they were able to invite a solidarity organisation to come to this UN meeting. I was lucky to be chosen to go – alone. I think I was only 19, I was young ... I had to make a speech and address this meeting; and that was a real scary thing.[18]

If Andre was slightly intimidated to be in this setting at such a young age, his participation in another delegation soon put this in perspective. Following the unbanning of anti-apartheid organisations in February 1990, City Group was invited to attend the first legal congresses of the PAC and the Black Consciousness Movement inside South Africa. Andre was selected as part of the City Group delegation. In addition to the conferences, their days were packed with meetings with a wide variety of anti-apartheid organisations:

> Every morning, every afternoon and every evening we had set ourselves up a whole range of meetings, meetings, meetings. It was absolutely amazing and one thing that struck me was you set up this meeting with some organisation, you expect to go there and meet, I don't know, what maybe we would think of as 'experienced' people, but they were *really* young. I was meeting people my age being General Secretary of this, or that, some committee, and that was pretty amazing.[19]

In London, critics of the Non-Stop Picket frequently dismissed it as being maintained by an unruly crop of naive youngsters who, although well meaning, were unwitting dupes of an international communist conspiracy against the government and South Africa (Westad 2005). When they travelled to Southern Africa, these young protesters often found themselves in dialogue with their direct peers who were trying to secure an alternative future for themselves without the inequalities, indignities, and violence of apartheid.

If some young picketers accepted the challenges and responsibility of representing the Group at high-level international meetings and conferences (Craggs 2014b; Craggs and Mahony 2014), others of their peers remember feeling daunted about the prospect of taking on more modest, everyday responsibilities on the picket. For example, Grace remembers that when she and two school friends joined the Picket, they resisted becoming Chief Stewards on their regular shifts (see Chapter 2) because "we were a bit younger [and] it meant dealing with the police, and it seemed such a big responsibility."[20] Irene also resisted this responsibility:

I hated it. It was the worst thing for me being given that. It was too much responsibility for me and I knew that at the time I think. I didn't like it. I didn't know how to handle the police. I didn't know how to handle people, the South Africans coming past and shouting about necklacing. I could handle it in the sense that I could shout back, but there was just a sense of responsibility for the whole Picket that I didn't want to have, and I didn't really know what to do, so I would only do it if I absolutely had to, if there was nobody else.[21]

Grace, however, acknowledged that despite her (and her friends') initial resistance, they eventually took on this role, and the responsibility that accompanied it, once they had been on the picket for a longer period of time:

To be honest I think it was when we saw other people who were like newer than us and we thought probably knew less than us, saw them taking on the role of Chief Steward, we thought well we could easily do that, and we've been here a long time, we know what the regulations are and the protocols.[22]

Whether in the role of Chief Steward, providing political leadership on a given picket shift and liaising with the police for a few hours, or being a guest of one of the South African liberation movements at the United Nations or one of their own conferences, young members of City Group were offered the opportunity to take on more responsibility than might usually be offered to people their age in other political groups. Not all young people welcomed this responsibility, but many found that they grew into it as their commitment to the Group developed and matured.

Organising youth

A key mechanism for providing young picketers with an opportunity to develop organisational and leadership skills amongst their peers was City Group's Youth and Students sub-group. This sub-group built support for the Non-Stop Picket amongst young people, touring the country to encourage students' unions and youth groups to support City Group's campaigns, pledge regular shifts on the Non-Stop Picket, and organise protests in their vicinity. The main focus of the sub-group's activities each year was organising a special protest on the Non-Stop Picket to commemorate the school students' uprising that erupted in Soweto on 16 June 1976.

During the spring term of 1987, City Group delegates spoke at students' union meetings up and down the country, mobilising in particular for the 14 March 'March for Mandela' demonstration, but more generally encouraging students to actively participate in anti-apartheid campaigning where they lived.[23] This speaking tour took in dozens of colleges around London, but also took City Group supporters south to Sussex University and north as far as Kilmarnock.

In March 1987, students occupied the main administration building at the London School of Economics (LSE) demanding that the college disinvest from companies profiting from apartheid in South Africa. Their occupation of Connaught House attempted to pressurize the LSE Court of Governors to withdraw £1.7 million of investments from Shell, BTR, BP, and Glaxo, all companies with significant business interests in South Africa.

Although City Group cannot claim responsibility for the LSE occupation against apartheid, there had clearly been a lot of contact between activists from the Non-Stop Picket and students at the LSE in the lead up to their action.[24] City Group played an active role in supporting the student occupation. When the students organised a rally in support of their occupation, City Group shared the platform with Jeremy Corbyn, Member of Parliament (MP), Diane Abbott, MP, Richard Balfe, Member of the European Parliament (MEP), (themselves all sponsors of the Non-Stop Picket) and a representative from the South West African Peoples' Organisation (SWAPO) from Namibia.

Ultimately, the students' occupation was defeated when the LSE secured a repossession order in the courts. Both Norma and David Kitson spoke at the picket of the court hearing organised by the students. When, after seven days in occupation, the students decided to vacate Connaught House of their own volition rather than resist eviction, 400 students marched spontaneously to the Non-Stop Picket of the South African Embassy at the other end of The Strand. While the close geographical proximity of the Picket to the LSE undoubtedly influenced the students' decision to march there, this also suggests that for youth and students interested in taking radical action against British links with apartheid, the Non-Stop Picket was a key focal point.

Despite City Group's commitment to providing a space within which young picketers could organise (semi-)autonomously, there could sometimes be moments of inter-generational incomprehension and miscommunication. Dominic, who later served as the Group's Youth & Students Officer, remembered one such early clash with Norma Kitson:

> I remember Norma getting cross with me once because in the youth and students meeting we were talking about what merch we could do like bookmarks or, you know ... Merchandise, like badges and matches, and then I think I suggested we could have burn down the Embassy on matches as a joke. I wasn't really suggesting it, but Norma got quite cross like how could you possibly? So I never said it again.[25]

Even within the sub-group itself, the different contexts around which members negotiated their activism could cause conflict. For Grace Livingstone, a school student, this was expressed through misunderstandings between herself and slightly older, unemployed activists who no longer appreciated the challenges of juggling her studies with her political engagement:

Because I [was] trying to do my homework in the youth and student sub-group ... Yeah, but the thing is I mean I was at school and doing my O-levels and he was like, and so he sort of said can you stop writing while I'm talking or something ... Yeah, I think he was the chair of the youth and students, and I was thinking well you know, I am a student and students sometimes need to do their homework.[26]

These misunderstandings and tensions did not prevent City Group from putting a lot of effort into mobilising students to participate in the Non-Stop Picket and other protests.

Youthful solidarity

In July 1988, the national Anti-Apartheid Movement (AAM) held a vast fundraising concert at Wembley Stadium to celebrate Nelson Mandela's 70th birthday in the company of George Michael, Whitney Houston, and other international stars (Thörn 2006). The country's "second-largest anti-apartheid benefit concert" of the year took place a month earlier at the Fridge in Brixton.[27] On 9 June, 1,400 people gathered in Brixton to hear The Wedding Present headline a benefit for City Group. All tickets for the gig sold out before the doors opened. The line-up also featured The Gargoyles (two former members of The Housemartins), Tony Smith from Radio London, and Chris McHallem (at that time playing Rod in the TV programme EastEnders) 'on the decks'. The bands wore 'Non-Stop Picket' t-shirts and encouraged their fans to join the Picket and, especially, to join the 'Surround the Embassy' protest planned for 16 June. While undoubtedly many fans just turned up for the music, this was a political benefit gig, and it was relatively difficult to avoid its political message.

Both the benefit gig and the high-profile attempt to 'Surround the Embassy' a week later had been organised by young picketers – members of City Group's Youth and Students sub-group.[28] Helen, one of the main organisers, celebrated their achievements:

We did a great benefit at the Fridge in Brixton with the Wedding Present and the Gargoyles and it was just ace. It was like really ace. We did loads of, you know, a lot of publicity for that. It was an amazing thing to actually feel, you know the Wedding Present were a big Indie band then, they had a big following, so to be able to secure them was a bit of a coup really. Yeah and we did good stuff. That was very good.[29]

The opportunity for this benefit initially arose out of a chance conversation on the picket in 1986 between Dominic and one of the band members from The Wedding Present. That two years later they capitalized on this conversation, when the band was at the height of their popularity, and drew in other well-known musicians and celebrities to support the event, demonstrates both the reputation and respect that the Non-Stop Picket commanded at that stage. As Dominic noted, "the Wedding Present were a big band at the time, I'd say second only to The Smiths in terms of like indie band rock gods."[30] The success of the benefit also demonstrates the practical event-management skills that young picketers had acquired through their political activity over the preceding two years.

Although the event was supported, in practical and political terms, by picketers of all ages, it was a cause of (mild) intergenerational dissonance, as Dominic recalled:

> Norma [Kitson] did the bookstall and she said to Carol [Brickley] that the event was like a sophisticated form of torture ... She wasn't going to pretend that she had fun at it.[31]

Previously, City Group's fundraising benefits had been quite modest affairs, mostly consisting of amateur musicians who were already associated with the Group in some shape or form. To line up one of the biggest indie bands of the time in a major music venue (rather than the auditorium of a town hall or the backroom of a pub) was a major coup for the Group. Unlike the AAM's Wembley extravaganza the following month (Thörn 2006), the City Group benefit was organised by young volunteers from the Group rather than professional event organisers.

While the funds raised on the night were modest compared to those collected by the Mandela birthday concert, they were still significant for City Group. Most of the profits were sent directly to South African liberation movements. In line with the Group's non-sectarian approach to solidarity, £1,000 was sent to the African National Congress (ANC), £1,000 to the Pan-Africanist Congress (PAC), and £500 to the (smaller) Black Consciousness Movement of Azania (BCMA). The cheques were sent with the clear indication that there were no conditions placed on what the money could be spent on (a coded consent that the funds could be used for military purposes). When we interviewed her, Helen Marsden made this point more explicitly:

> We raised some good money and it was nice to be able to say that we'd raised this amount of money and we were going to split it equally between certain parties and I do remember having a conversation saying and if they want to spend it on guns, we're going to let them spend it on guns, it's their money.[32]

The additional funds sent to the PAC and BCMA as a result of this gig were appreciated and needed (in the context of City Group's strained relationship with the London ANC (see Chapter 2), it is less clear whether the ANC actually ever cashed the cheque sent to them – with more established international funding networks, they could probably afford to turn down money from City Group (Ellis 2013; Herbstein 2004)). The fundraiser was a major achievement for City Group and demonstrated the reputation that the Non-Stop Picket had gathered over the previous two years.

Soweto Day commemorations

Every year, City Group commemorated the 1976 school students' uprising in Soweto. On 16 June 1988, the Group attempted not just to swell its continuous protest outside the Embassy, but also to entirely surround South Africa House. It *just about* managed it with a chain of people marching around the block that the Embassy occupied and more or less encircling it. The 'surround the Embassy' theme was so popular that another attempt was made the following June (Figure 7.1).

Understandably, commemoration of the Soweto school students' uprising became the focus for City Group's efforts to mobilise and activate more young people to anti-apartheid solidarity. However, the way in which the story of the Soweto uprising was retold suggests a lot about how City Group understood the role of youth in anti-apartheid campaigning (Selbin 2010). The leaflet for the 1989 Surround the Embassy protest opened with the following text:

> At 7am on Wednesday 16 June 1976, thousands of black students assembled for a march through Soweto in South Africa in protest at Bantu Education – the enforced teaching of Afrikaans, the racist wite [sic] minority's oppressor language. The students began to march to the Orlando Stadium when the 'Security' Forces opened fire with guns and

Figure 7.1 Young picketers at the Surround the Embassy protest, 16 June 1988.
Source: Photographer: Jon Kempster.

grenades on a crowd armed with banners, placards and slogans. Hector Peterson, 13 years old, was the first to be murdered.

The anger of the demonstrators then exploded. They built barricades, stoned the police, put government buildings and property to flame, they burnt down the offices of the Bantu administration. The uprisings continued for three days and spread up and down the country, while police massacred more than 1000 youth.[33]

This is a fairly standard representation of the school students' protests (Thörn 2006: 159). However, rather than victims, it presents the youth of Soweto as active agents of resistance to apartheid. They fought back and escalated the rebellion. This resistance was portrayed as inspirational for British youth, who were encouraged to contribute through their own direct action, rather than more passive humanitarian sympathy.

The next paragraph of the flier is equally significant, both in terms of how it presents the Soweto youth, and for its presentation of City Group's approach to solidarity:

Many thousands of youth secretly left the country in the following months to join the armed struggle with APLA and Umkhonto we Sizwe. Steve Biko was murdered for his part in organising the demonstration. Zephania Mothopeng, President of the Pan Africanist Congress, has only just recently been released from prison for his part in organising the demonstration.[34]

City Group and the Non-Stop Picket were committed to offering 'non-sectarian' solidarity to all political tendencies fighting apartheid in South Africa. That commitment is exemplified in this text. APLA, the military wing of the Pan-Africanist Congress, are mentioned before the (better-known) armed wing of the ANC. The significant role of leaders from the Black Consciousness and Pan-Africanist traditions in inspiring the uprising is unambiguously acknowledged through the naming of Biko and Mothopeng.

Over the years, several attempts were made to surround the Embassy. The June 1988 attempt was probably the most successful (although whether it was actually 'surrounded' is open to debate). That Thursday evening, 1,500 people gathered outside South Africa House.[35] The numbers were undoubtedly swelled by the publicity that had accompanied the previous week's benefit gig. In the spirit of the Soweto uprising, as the rally opened, it was chaired by Daniel, an active member of City Group's Youth and Student sub-group, then aged 15. He introduced a variety of speakers that evening including Zolile Keke from the PAC, Norma Kitson, Peter Tatchell, and John Mitchell (the General Secretary of the Irish trade union IDATU). In between these speakers, and enlivened by the music of the Batucada Mandela samba troupe and ska from the Horns of Jericho, a crowd of several hundred people marched round the Embassy three times before flowers were laid on its gates, and 600 black balloons were released. At this point, the Metropolitan Police's Territorial Support Group was brought in and used to move protesters from in front of the embassy gates. They forcefully pushed and dragged peaceful protesters out of the way. Some were physically picked up and thrown over the crowd-control barriers, and one was dragged backwards by the placard strung round her neck. Four key City Group activists were arrested in the process, with a fifth being arrested later in the evening. John Mitchell said at the time: "The savagery of the police attack on the peaceful picket was the worst I have witnessed since I saw Marcos' troops batter workers in the Philippines."[36] At 10:30 p.m. that evening, there were still over 300 protesters on the picket.

For Susan, who was barely into her teens at the time, the practice of adorning the embassy gates with commemorative flowers was particularly poignant (as well as being an opportunity for youthful rebellion that felt very empowering):

> And I remember being very excited because we were allowed to climb up the gates to put the flowers up there, and there would always be a few adult comrades guarding us or giving us a foot up or yeah protecting us. And I remember even at times holding the police back from us really from doing that job. And it was just an opportunity to rebel and to just to have a little statement against authority that we were going to do that. And felt very privileged to have been a part of that because obviously remembering the children of Soweto and being a child at the time, this made you realize that just battling through a few police who obviously in those days didn't wear riot gear and putting flowers on the Embassy was such a small task compared to the children who were throwing stones and rocks and fighting for their rights in South Africa.[37]

The following year's Soweto commemoration rally was a smaller affair, attracting a crowd of around 400 people.[38] On this occasion, City Group could not even pretend to have surrounded the Embassy (although that had been its intention). However, once again, the bulk of the crowd did march around the perimeter of the Embassy before 17 volunteers[39] lined up in front of the embassy gates and turned round the generic placards we were holding to spell out "Free the Upington 26" in solidarity with a group of detainees in South Africa, some of who were facing the death penalty for their part in an anti-apartheid protest in the township (Durbach 1999). Under cover of this visual (decoy) protest, three activists chained themselves to the embassy gates before being quickly cut loose by the police. There were no arrests.

The Soweto commemoration rallies on the Non-Stop Picket provided a focus for British youth to show their solidarity with young people resisting apartheid in South Africa. In remembering the youth of Soweto, the Surround the Embassy protests provided City Group with the opportunity to publicly practise its non-sectarian approach to anti-apartheid solidarity and acknowledge the achievements of the Black Consciousness tradition and the PAC. But, even more so than the habitual practices of the Non-Stop Picket, it created an opportunity for young solidarity activists to (been seen to) take responsibility for organising a major protest. These youth-centred protests helped consolidate the strong interpersonal bonds between young picketers.

Growing up together

The Non-Stop Picket created an atmosphere where young people's political opinions and motivations were taken seriously. The protest was open, welcoming, and inclusive – it encouraged active participation. Keeping the Picket going, as a continuous protest, provided many opportunities for young people (including pre-teen children) to take on various responsibilities. In taking on these responsibilities (and being trusted by older people to do them), young picketers developed new skills, but also learned more about themselves.

For young people, being on the picket was both serious and fun. They were taking action for a political cause that they believed was important. In the process, they were taken seriously by adults and given opportunities to develop new skills and confidence. Although young people could sometimes feel intimidated at first by the responsibilities entrusted to them, the same people also reflected that, as teenagers, they often had an inflated sense of their own maturity and capability. This is a point that was made by both Grace and Georgina in their interviews:

> The thing is, when you're 15 you feel quite grown up, you don't think I'm really young, although maybe other people think you're young, but you feel completely grown up, so that didn't really occur to us.[40]

> I was young I think in a way probably that gives you a false sense of security anyway. Because I think that when you're that young, I mean I was 16 when I started, 16 going on 17 when I started on the Picket and I think you kind of just take what comes to you and you don't question it particularly in that way.[41]

There is no question that these young picketers took their politics incredibly seriously, but time spent on the Non-Stop Picket, with other politically engaged young people (and older campaigners with a wealth of stories and experience), was fun and exhilarating. Andre, who eventually moved from the Netherlands to be part of the Picket full-time, remembered:

> It was very exciting. I mean of course I loved it, I was part of it, I wanted to be part of it. It felt like you had a purpose, you had a grand thing. Yeah, so I think I got a lot of strength from it, I mean we can go onto – I mean

I obviously have to go onto that about how it shaped loads of lives and we learned so much and gave us untold leadership skills and lots of other things.[42]

For Grace, part of the appeal of the Picket was that it allowed her to explore her emerging political beliefs and becoming political in the company of other teenagers who were going through the same process:

> So the main thing, as I was saying, we were sort of like young people just becoming politicised, and the Picket was important because it got you together with other people who sort of felt the same way. I mean obviously you made friends as well and that was really nice, but it was mainly that sense of like working together with people who felt the same way, making your voice heard. Yeah, it was sort of making your presence felt and like express, it was a way of expressing your outrage and just making your voice heard really.[43]

Through their shared commitment to the Non-Stop Picket and the anti-apartheid cause, a diverse group of young people enjoyed their youth and became adults together, as well as finding ways to deal with (and support each other through) the everyday challenges and dramas of late adolescence. Anti-apartheid campaigning was not a backdrop to these young people's lives; they grew up *through* their political engagement. Time spent on the picket played a profound role in enabling participants to cope and live through troubling (even traumatic) life events. For example, Daniel suggested:

> There were a lot of people who had similar kind of strange families or kind of funny home situations that it felt more comfortable on the picket, that's kind of how I felt ... And there was also a sense of everybody having this common experience, everybody being in the same situation and whatever crisis or drama was current at that point that people were working together to try and find a way round things to try and come up with ways of responding to situations and trying to kind of control the situation and that there was a word that I've heard a lot repeated by people since is that you learned a hell of a lot about the nature of solidarity there. Really it was kind of like you learned that there were people there that you could really trust as well. You could trust them politically and you could trust them personally.[44]

The practice of regularly standing together in front of the Embassy nurtured strong bonds of social solidarity amongst the Group. Of course, time spent on the picket, or in the Group's office, or, for that matter, a police cell was never just 'political'. There was a strong social component to making the hours pass on the picket, and young picketers, in particular, developed a lively social life around their picket shifts and other political commitments. For Deirdre, who joined the Picket shortly after moving to London from Ireland on her own, this was particularly valuable:

> So yeah, apart from this political message it was really that social side. I mean I was very young, I was 19, of course I was going to want to be around people, and I certainly was a very social person. I think so yeah, it was important, and I fell in love on the picket, all of that. I suppose I fell in love on the picket and with the Picket. So that certainly, I think if hadn't met those people, like yourself [Gavin] and others, and created and maintained very strong friendships well after the Picket stopped during the years that I was still in London, I think I would have still stayed. I think my political conviction was such that I, if I only thought they were all right, I think I still would have stayed involved. But the fact that there was the sociability, there was the people that did want to have a bit of a laugh as well as, you know, because we were young, why wouldn't we want to have a laugh too? I mean it was already extra-ordinary the amount of time we gave to this campaign – that was very important, it was certainly a bonus, it was a lovely bonus.[45]

The close web of friendships that developed between young picketers had many implications for their lives at the time. The Picket provided a safe and supportive environment in which young picketers could explore their identity and sense of self (in several cases, their sexuality) and try out new ways of being. Irene gave a very clear sense of how the first year of the Picket coincided with significant milestones in her transition to adulthood (and an associated period of reflection on the person she wanted to be).

> I moved out of my parents' house just after my 16th birthday, which was in February of the year [the Picket] started, so I was living alone for five months I think, before I joined, in a bedsit and it felt like, I suppose that year for me was turning 16, moving out, joined the Picket, did my exams, left school. It felt like the whole thing was a big step towards independence I suppose, and so the Picket, it did represent that, it did represent kind of striking out against your past to some extent and saying, right, I'm not that person anymore, I'm this person, and trying to find out I suppose who you want to be.[46]

Ultimately, however, as the buzz that developed around the June 1998 benefit gig at the Fridge in Brixton demonstrates, young people were also drawn to the Non-Stop Picket because it was '*the* place to be'.

Intergenerational interactions

People of all ages participated in the Non-Stop Picket and worked together, side by side, to maintain the Picket as a continuous protest. This section of the chapter considers the ways in which the Picket provided opportunities for intergenerational interaction and friendship, as well as moments of incomprehension and miscommunication across the generations (Hopkins and Pain 2007; Vanderbeck 2007).

If some of City Group's campaigns were led by youth and students, there were a small number of campaigns that never really engaged and enthused the younger picketers (even if they appreciated their political importance). The Justice for Kitson Campaign was one such campaign.[47] Although City Group supported the campaign, it was largely driven by older British communists who had worked with David Kitson in the Hornsey branch of the Communist Party in the 1950s, along with supportive members of his union and the Ruskin College community (where David had been promised a fellowship on his release from gaol). The Justice for Kitson Campaign fought for Norma and David's reinstatement to the ANC, for the reinstatement of the union's funding for David's post at Ruskin, and sought to establish a fund to support David and others in similar situations.

To an extent, although they were generally supportive of the campaign's aims, the hard slog of winning support for David through the trade unions was not an attractive form of campaigning for many of the young activists who sustained the Non-Stop Picket at the time. Conversely, few of David's old comrades from his union and the Communist Party could sustain long shifts on the pavement outside the Embassy. There was something of an age-defined division of labour between these complementary campaigns.

As Irene noted, for teenagers, even a few years age difference can seem enormous. This could be both exciting and intimidating:

> People seemed very old to me! I met the younger people later and I think we became very close because we were uniquely young at the time, so I think people who were in their mid-twenties, I remember thinking, God I'm hanging around with real adults, that seemed really exciting, to be hanging out with old people![48]

There are two points to be made here: First, the age difference between picketers at either end of the 'youth' category could be experienced (particularly by the youngest picketers) as dauntingly significant. Second, through propinquity on the picket and joint campaigning, the Non-Stop Picket provided an opportunity for many young picketers to engage with adults (other than their parents and teachers) as equals for the first time. Irene continued to explain how she experienced this:

> I think it was my first introduction to people above my own age. I never mixed with the friends of my parents, which I think is where you'd normally come across adult relationships, I never, ever did that … and I think it was very, very good for not patronising you. It really did. People were treated as equal no matter what their background or their age or anything else and to be treated alongside people who I thought was really old, the 25-year-olds and the 30-year-olds and obviously older than that. You'd suddenly be standing next to someone who was 60 and I'd never had that. I'd only ever had school friends.[49]

Like Irene, many young picketers developed a real respect for adult picketers like Norma Kitson, who they felt had lived 'real political lives'. These older comrades, who had dedicated many decades to campaigning for progressive and radical causes, were an inspiration for young(er) picketers (of all ages). Rene Waller, a lifelong communist who continued to serve as City Group's Membership Secretary and take part in direct actions (such as occupations of the South African Airways offices) well into her seventies, was fondly remembered by many interviewees.[50] Some, like Francis, remembered her as a friend and mentor:

> Rene Waller (1913–1999) was a great friend of mine. I kept in touch with her right to the end. Rene showed me that you will be most effective as a militant revolutionary if you remain polite, calm, disciplined and efficient.[51]

For Helen Marsden, who had dropped out of sixth-form studies in order to dedicate her time to anti-apartheid campaigning, one of the unexpected pleasures of life on the Non-Stop Picket was the opportunity to develop genuinely close friendships with women who were many decades older than her:

> Marge was like, we were really, really good friends, and we used to go to the cinema a lot together. And so just the very thought of entertaining, having friends who were pensioners was unbelievable. But I'd say, yeah, most of my friends I think were older than me. So, yeah, I probably, if I'd have stayed in a sixth form, go to university, go to work sort of line of things, I wouldn't have had, developed those friendships.[52]

These young picketers, who were in their teens in the late 1980s, were in their mid-to-late forties by the time they were interviewed for our research. In this context, it is inevitable that we were unable to record too many stories from the (considerably) older protesters with whom they developed intergenerational friendships. Nevertheless, the following comments from Simon (who was in his early thirties when he joined the Picket and became close friends with many fellow picketers who were nearly half his age at the time) give an 'adult' perspective on these interactions:

> Most of the people that I'm even still in touch with are quite appreciably younger than me ... the Picket was full of people from so many different organisations, backgrounds, etc. I don't think there would have been a way for me or any of us really to get involved with such a mixed bag of individuals. We were, everyone was very different and had their own spin on things and it's like I said, it was all part of a political education for me that I felt was really lacking before I got involved on the Picket. And it was people from all, across all ages as well and obviously most of the people that were involved on the Picket were obviously younger than me and they taught me a lot and I have no shame in saying that either.[53]

As Simon makes clear, it was not just teenagers and other youths who felt they benefited from and grew through the intergenerational interactions and friendships that they developed on the Non-Stop Picket. City Group's solidarity work held together protesters from many different generations. Although some activities effectively became differentiated between age cohorts, and there could often be misunderstandings and tensions between different generations of picketers, habitually, people of different ages worked together, got to know each other, and learned political lessons from each other.

Youthfulness beyond the youth

This chapter has argued that young activists' political commitments are always entangled with the everyday politics of youth; that (in the context of the Non-Stop Picket) to practise solidarity was also to develop competences and resources that contributed to the practice of growing up. Although this chapter has primarily focussed on the experiences of teenagers and young adults, we also argue that 'youthfulness' and practices of 'growing up' are relational and not age specific. For example, time spent on the picket (with or without their parents) enabled pre-teen girls like Susan[54] to hang out with older teenagers and young women, emulating their style and their political commitment (Taft 2011):

> We would always be friendly and sociable to other picketers, and I suppose there were, as young girls, women picketers who we were sort of like wowed by and, you know, we just thought they were so cool and pretty and revolutionary strong women, good role models. And they also reminded us of our pop stars at the time, as I remember Georgina, you know, she just reminded me of Madonna. So they were really cool. And we weren't in their circle. We were not teenagers at the time, we were younger, but we liked their attention and we liked being around them. And we socialised [with them] in a sense that we could all sing the songs together and we would all have a giggle together.[55]

In this way, time spent with teenage picketers allowed those children who regularly visited the Non-Stop Picket to model and anticipate a youthfulness that they would soon grow into. Time on the picket also allowed older picketers to be 'youthful' in various ways. A number of picketers who were in their thirties when they encountered the Picket described how their involvement with the social and political life of the Non-Stop Picket gave them opportunities *both* to relive their youth *and* to 'really grow up'. In this context, let us pick up Simon's story. Even though he had previously been active in the Labour Party and his trade union, Simon argued that he also matured politically through his engagement with the Picket:

> I think I was a bit naïve before I joined the Picket political wise and I didn't really know what I wanted to do or who I wanted to be with, and I think the Picket really helped me.[56]

More significantly, Simon (who was going through a divorce) reflected that, like many of the teenage picketers, he was going through a difficult and disruptive life transition, and his commitment to the politics of the Picket and the comradeship of his fellow picketers enabled him to cope with this.

Penelope, who had recently graduated as a mature student and, by her own admission, was a little lost for what to do next, made some very similar observations about her experience of living through a liminal period of personal uncertainty:

> All of these encounters were mind-bogglingly educational for me in their fresh original outlooks and ideas, and challenged and broadened my perspective and made me less gullible, and more accepting. Of course for all of us who participated the wonder of it was that life went on for all of us. We all had problems of a practical, financial, health or emotional nature. Family, work, other commitments pulling on our commitment and having to be addressed in some degree ... People had unwanted pregnancies, destructive relationships, housing problems etc. etc. etc.[57]

Despite these personal difficulties and personal traumas, Penelope recognised that "all those friendships were most stimulating, changed me forever, and kept me young."[58] The experience of 'growing up' on the Non-Stop Picket might have been most intense for teenage picketers, but as the testimonies of people like Simon and Penelope attest, London's continuous anti-apartheid protest also provided opportunities for all ages to experience the exuberance and joy of 'youthful' activism. Picketers of all ages benefited from interactions across generational boundaries, but those people who (like the teenage activists) were also experiencing a significant life-course transition were often those who enjoyed the social life of the Picket, the mutual support it offered for those finding their way through the new challenges life threw at them, and remember those years as a particularly intense period in their personal (not just political) lives. In the next chapter, we consider how picketers coped when the Non-Stop Picket came to an end.

Notes

1. Interview with Georgina Lansbury.
2. Interview with Andre Schott.
3. Interview with James Godfrey.
4. Interview with Ben Evans.
5. Interview with Gary Lowe.
6. Interview with Nicole Aebi.
7. Interview with Rikke Nielsen.
8. Interview with Ben Evans.
9. Ibid.
10. Ibid (emphasis added).
11. Interview with Mark Farmaner.
12. Interview with Andre Schott.
13. Interview with Grace Livingstone.
14. City Group (1986), 'Non-Stop Picket until Mandela is Free' (November/December). This general leaflet advertised an evening rally on the Non-Stop Picket on Wednesday 10 December to mark Namibian Women's Day, the day school on Sunday 7 December, and fundraising concert 'Music for Mandela' at Islington Town Hall on Saturday 13 December. In itself, this leaflet suggests the pace at which being 'non-stop against apartheid' was lived.
15. City Group (1988), *Picketer's News*, 1 (20 March).
16. Interview with James Godfrey.
17. Interview with Charine John.
18. Interview with Andre Schott.
19. Ibid.
20. Interview with Grace Livingstone.
21. Interview with Irene Minczer.
22. Interview with Grace Livingstone.
23. City Group (1987), 'We are here until Mandela is Free. Demonstrate 14 March 1987', Non-Stop News, 16 (23 January), p. 1, 4; S. Dewhurst (1987), 'City AA goes on tour', *Fight Racism! Fight Imperialism!* March, p. 3.
24. A. Burrows and L. Morgan (1987), 'LSE students occupy for disinvestment', *Fight Racism! Fight Imperialism!* March, p. 3; Lorna and Ann (1987), 'Young people say: We want Mandela free', *Non-Stop News*, Anniversary Issue, p. 7.
25. Interview with Dominic Thackray.
26. Interview with Grace Livingstone.
27. City Group (1988), 'Pressies for City AA', *Picketers News*, 4 (10 June), p. 2; D. Jewesbury and R Roques (1988), 'Brixton! Soweto!', *Non-Stop Against Apartheid*, 29 (July), p. 8.
28. City Group (1988), *Non-Stop Against Apartheid*, 28 (June), contained a box advert for the 'Surround the Embassy' protest (p. 3), and a centre page feature 'Youth Under Apartheid' by Mike and Barry (pp. 4–5).
29. Interview with Helen Marsden.
30. Interview with Dominic Thackray.
31. Ibid.
32. Interview with Helen Marsden. In contrast, Fieldhouse (2005: 279) stresses that the AAM was sometimes a little ambiguous in its support for the ANC's armed struggle for fear of alienating potentially supporters in the UK.
33. City Group (1989), 'Surround the Racist Embassy', A5 leaflet (May/June), p. 2.

34 Ibid.
35 M. Burgess and D. Thackray (1988), 'Embassy Surrounded', *Non-Stop Against Apartheid*, 29 (July), pg. 8.
36 Ibid.
37 Interview with Susan Yaffe.
38 D. Thackray (1989), 'Remember Soweto! 400 surround the racist embassy', *Non-Stop Against Apartheid*, 35 (July/August), p. 7.
39 Including Gavin.
40 Interview with Grace Livingstone.
41 Interview with Georgina Lansbury.
42 Interview with Andre Schott.
43 Interview with Grace Livingstone.
44 Interview with Daniel Jewesbury.
45 Interview with Deirdre Healy.
46 Interview with Irene Minczer.
47 *Justice for Kitson Campaign* pamphlet (1988), London.
48 Interview with Irene Minczer.
49 Ibid.
50 R. Waller (1993), 'Memories of a still-fighting communist', *Fight Racism! Fight Imperialism!*, June/July, p. 12.
51 Interview with Francis Squire.
52 Interview with Helen Marsden.
53 Interview with Simon Murray.
54 Susan was 11 when the Non-Stop Picket started.
55 Interview with Susan Yaffe.
56 Interview with Simon Murray.
57 Interview with Penelope Reynolds.
58 Ibid.

8 'Until Mandela is free …'

It was a great day obviously. It was amazing because obviously we'd been working for this for so long. I remember the day they announced [Mandela's] release even, coming to the Picket and holding up *The Evening Standard*, Mandela to be released, I think it was a real joy. And Danny coming with this little, he always had this little Walkman thing with Free Nelson Mandela, playing it over the megaphone.[1]

Yes, it was big. It wasn't like the Picket almost, you know, there were so many people who had gone to Trafalgar Square, including the AAM who had their own platform and were trying to make sure they were the only people who were interviewed. It was exciting … I felt that City Group had been vindicated.[2]

When Mandela from gaol after 27 years imprisonment those who had maintained a non-stop picket outside the South African Embassy for so long wanted to celebrate. They felt a sense of achievement and that their determination had been vidicated. The primary demand of the Non-Stop Picket was Mandela's unconditional release. So, when he was finally released in February 1990, the Picket had achieved its main aim and had to come to an end. Although City Group continued campaigning until apartheid ended in 1994, with the first non-racial elections in South Africa, the majority of picketers quickly disengaged from its work once the Non-Stop Picket was over. Mandela's release (and the subsequent last day of the Picket) was celebrated as a victory; but for many participants, the abrupt end of the Picket felt like a loss. The protest that had become the focus of their lives over the previous four years was gone, and the close comradeship they had developed there was threatened. This chapter examines how the Non-Stop Picket ended and how City Group's activists reacted to this. It examines some of the ways in which City Group continued campaigning against apartheid after the Non-Stop Picket. Finally, it also sets out some of the ways in which former picketers reflected on the post-apartheid settlement in South Africa.

Struggling to be non-stop

Many of the stories told in this book recount tales of daring and defiance by (mostly) young activists; some celebrate political victories. It is important to remember those stories, but any attempt to seriously tell the history of a protest movement also has to examine the more mundane, possibly boring, aspects of political activism (Chari and Donner 2010; Davies 2012). Protests seldom happen spontaneously; they need organising. To make a protest happen, a whole range of people, ideas, and resources need to be mobilised. More than this, even before leaflets can be written, publicity distributed, props assembled, and potential participants contacted, the political situation needs to be analysed and assessed, strategic priorities need to be reviewed, and tactical decisions need to be reached. In many ways, the final months of the Non-Stop Picket were exhausting, and the protest was becoming harder to sustain. Beyond their dogged determination to fulfil their promise to maintain a continuous protest until Mandela was free, the Group's analysis of the situation in South Africa undoubtedly contributed to the resilience of City Group's core group of activists.

Over the weekend of 23/24 September 1989, City Group held its Annual General Meeting (AGM).[3] Although no one could know for sure at the time, it was to be the last AGM the Group held before the end of its non-stop picket. Although the meeting ended up agreeing a busy programme of action for the year ahead, it took place at the end of a tense and disorientating time for City Group. Following de Klerk's ascension to the leadership of the National Party earlier in the summer, the situation in South Africa was moving fast. City Group had been responding dynamically to these events and had just pulled off one of its most dramatic protests to date, when hundreds of the Group's supporters blocked the road outside the Embassy in protest at the white-only election in South Africa on 6 September 1989 (see Chapter 6). Simultaneously, time was passing slower and slower for many City Group activists, as it was becoming increasingly difficult to sustain the Non-Stop Picket as a continuous protest. With some shifts on the picket's rota becoming increasingly difficult to cover, and growing numbers of people failing to turn up for their regular shifts (including members of the Group's committee), a potent combination of political tension and interpersonal frustration was growing within the Group.

Carol Brickley, City Group's Convenor, opened the AGM with a speech that analysed recent events in South Africa, reviewed the tasks of a solidarity organisation in Britain, and sought to unite the Group's membership behind a programme of action for the coming period. This session of the meeting was attended by Rodwell Mzontane, the Chief Representative of the Pan-Africanist Congress (PAC) of Azania in the UK at the time. He contributed the perspectives of the PAC to the debate. Carol opened her speech with the following observations:

It has been a year in South Africa where the struggle has again escalated. Events throughout the year have reminded us that the militancy that was demonstrated in the country in '84/'85 is still there, still exists, and is still ready to come to the fore. The state of emergency and the repression that followed the '84/'85 uprising has failed to quell the aspirations of the black people in South Africa.

This year has been marked by a major offensive by the imperialists in relation to Southern Africa. The situation within the country and within the region has to be viewed as part of that international imperialist offensive, in particular by US and British imperialism. Thatcher and the Reagan/Bush administration are crusading throughout the world to destroy the influence of socialism and communism. That shouldn't be seen just as a question of how much they loathe the notion of democracy. It is not a purely ideological question. It is not simply a hatred of socialism that motivates them. What is going on in Southern Africa, and indeed throughout the world, is a drive to establish new markets for capitalist exploitation without which imperialism cannot survive.... So, imperialism is searching for a solution to the question of apartheid, not on behalf of the black people of South Africa but on behalf of its own moneybags, on behalf of its own profits.[4]

The clarity and power of this anti-imperialist analysis is not surprising – Carol Brickley was a long-standing member of the Revolutionary Communist Group (RCG). Although City Group was never simply a front for the RCG (as many in the Anti-Apartheid Movement, and more broadly on the British Left, suggested at the time) (Fieldhouse 2005), they, in alliance with Norma and David Kitson, ensured that City Group was guided by anti-imperialist perspectives. It is these politics, as much as the Group's dynamic commitment to street-based politics and direct action, that differentiated City Group from the mainstream Anti-Apartheid Movement and the anti-apartheid solidarity of other political tendencies on the British far Left (Brown 2018).

In the context of that AGM, however, Carol's analysis served another purpose too. In the weeks and months leading up to the meeting, a group of young picketers[5] had become increasingly frustrated with the behaviour of some members of the RCG and – for right or wrong – projected many of the problems of the Non-Stop Picket on to that group as a whole. Undoubtedly, anti-communist sentiments motivated some of those critiques, but many raising those concerns understood themselves as anti-capitalists in some shape or form. Carol's speech spoke to them. The message was clear – if you are committed to the liberation of the black majority in South Africa, don't allow 'imperialism' to force a wedge between you and your communist comrades. It was an effective strategy. Although the AGM did not resolve all the tensions within the Group, it did serve to unite the majority of City Group's core activists behind a shared programme of action.

Carol concluded her speech with this call to action:

> City AA will have a hard task in 1990. Most British people will be easily convinced that apartheid is ending and that it is time to furl our banners. We have a different view. Solidarity with the liberation struggle and our capacity to undermine imperialism in its heartland will be more important in the coming year, as imperialism strives to rescue South Africa for itself at the eleventh hour... We have to go forward to a better organised, better politically focused City Group which is going to prove more than a thorn in apartheid's side... let us start by making it into a spear in the side of imperialism.[6]

With that challenge ringing in their ears, the 50 activists gathered in that room went on to debate and commit themselves to a punishing schedule of campaigning. They planned a renewed campaign for a consumer boycott of South African goods (Jackson 2004), committed to continued active protests in support of the sports boycott (Booth 2003), and pledged to hold a national demonstration in support of the Upington 14 (Durbach 1999)[7] in March 1990. Even though it is clear in hindsight that many of the activists in the room were experiencing burnout (Brown and Pickerill 2009; Cox 2010) at the time, after sustaining the Non-Stop Picket continuously for more than three years, the Group renewed its commitment to being 'non-stop against apartheid'. One aspect of the material presented in this chapter is to acknowledge some of the human costs that that renewed commitment entailed. Even so, the September 1989 AGM, and Carol Brickley's speech at it, was successful in re-assembling the different constituencies within City Group to continue working together in radical solidarity with those resisting apartheid in South Africa as part of a broader anti-imperialist worldview (Brickley 1985; Brickley et al. 1986).

Celebrating Mandela's release

On 2 February 1990, at the opening of the new (whites-only) Parliament in South Africa, President FW De Klerk announced the unbanning of the African National Congress (ANC), the PAC and the South African Communist Party (Lester 1998). He also announced that Nelson Mandela was soon to be released from goal.

From the release of Walter Sisulu and other leading prisoners the previous October, it had become increasingly likely that Mandela's release was going to be forthcoming.[8] For 46 months, the Non-Stop Picket had remained outside the South African Embassy in London with a stated commitment to stay there until Mandela was released from prison. With that event finally on the horizon, City Group suddenly had to prepare for the end of this long-running protest and make plans for a new phase of solidarity activism.[9]

On the day of the announcement, City Group held a celebration rally outside the South African Embassy. In typical fashion, the police did not allow the celebrations to go unscathed – during the day they arrested one picketer, Leigh, for attempting to tie the black, green, and gold colours of the ANC to the gates of the South African Embassy. The relationship of the anti-apartheid protesters to the space outside the Embassy had always been highly contested, and it remained so until the very end of apartheid, with the first non-racial election in South Africa in April 1994.

On 11 February 1990, in anticipation of Mandela's release from gaol, thousands of people converged on the South African Embassy to celebrate the occasion.[10] Although only a small proportion of the crowd who gathered on the day of his release had ever directly participated in the Non-Stop Picket, its continuous presence over the preceding four years had clearly seeped into Londoners' consciousness. That day, the pavement outside the South African Embassy became the place to be.

Something of the atmosphere that day was captured by Tony Benn in his published diary *The End of an Era* (1994: 585–6), where he recalled being phoned that morning and invited to attend the celebrations - carefully noting that, by then, City Group had maintained the Non-Stop Picket for 1,395 days and nights. He presented a picture of a "marvellous event" filled with singing, dancing and hugs; comparing the atmosphere to the celebrations at the end of the Second World War.

One former City Group activist, Francis, gave his own account of the day (including an interesting observation about Tony Benn's reception by some in the crowd):

> I arrived fairly early in the morning when it was still very quiet. Other City Group people began to arrive, and it started like a typical City Group rally. ANC people were there too and setting up a stage and amplification equipment. They were actually very friendly and appreciated our presence. As more and more young black South Africans arrived, it became more and more and ANC event. "Free Nelson Mandela" and Miriam Makeba records were sounded through the amplifiers. Speeches were made, by various people, including Tony Benn, who was heckled by some about the 1967 Namibian uranium incident.[11] The young South Africans started to sing, and then they were toyi toying. It was wonderful! I felt as though I was at a real South African celebration and I was too! The general atmosphere was electrifying although I did encounter a small amount of hostility when I took a break from singing and dancing to distribute leaflets to the crowd. "I didn't come here for this!" snarled an indignant woman. But most people were joyful about the great news and full of admiration for what we'd done.[12]

For those who had been a regular part of the Non-Stop Picket, this was a day of celebration. It was the culmination of what they had been campaigning for and provided a genuine sense of victory (Figure 8.1). When asked if there were specific days on the picket that they particularly remembered, many former picketers identified this day:

178 *'Until Mandela is free ...'*

Figure 8.1 Picketers celebrate Mandela's release, 11 February 1990. Source: Photographer: Jon Kempster.

> The demo when Nelson Mandela was released and we just shut London down ... I lost my voice for three days. I still get emotional when I see him on TV.[13]

> And of course the day Mandela was released. The incredible swell of people, completely taking the police by surprise until we took over Trafalgar Square ... A wonderful day. It started out quite small with a sense of "is it really happening" I think. I can't quite remember who was on the megaphone when the announcement came over (Richard perhaps?). The crowds grew and grew, the police couldn't cope. There was impromptu partying and music making on the steps of [St Martin-in-the-Fields church], and just hundreds of people.[14]

Many young people from across the world had been part of the Non-Stop Picket over the previous four years. For those who had left London by the time of Mandela's release, not being present on the picket on the day of his release was (and still is) a source of hurt and regret. Beth's comments give a clear sense of this:

> [I] was already back in Brazil ... I felt as if I had run miles, to fall some inches from the finish line ... was awful not to be there.[15]

Francis's comments [above] demonstrate a generosity of spirit and a recognition that that day was a day of celebration not just for City Group activists, but for exiled ANC members and other progressive South Africans in the UK, as well as a far-wider layer of people who were against apartheid. If relations between the ANC and City Group had been tense for many years (see Chapter 2 for an overview), on 11 February 1990 (despite some tense exchanges between individuals), a spirit of temporary cooperation more or less held. Nevertheless, many regular picketers felt quite territorial about the pavement and pointedly questioned where these hundreds of revellers had been for the previous four years.

Although City Group had always stated that they would maintain the Non-Stop Picket until Mandela's release, the protest did not end the day he walked free from jail. The Non-Stop Picket kept going for another fortnight. In part, this extension occurred out of concern that the apartheid regime were not acting in good faith and that restrictions might be placed on Mandela's freedom, or that Right-wing paramilitaries might make an attempt on his life. The delay also gave City Group time to plan their strategy for campaigning post-picket and allowed the Group to come together to celebrate their achievements, closing down the Picket on their own terms. The final day of the Picket provoked a far more mixed emotional response from picketers.

Ending the Non-Stop Picket

On Saturday 24 February 1990, after 1,408 days and nights, City Group ended its Non-Stop Picket. When the Picket started, City Group pledged to maintain the Non-Stop Picket until Nelson Mandela was released from jail (even though that task seemed daunting). To echo a comment by Carol Brickley that we quoted in Chapter 2:

> You don't say I am doing this because it's possible, because frankly it didn't seem altogether possible. But the value of it was in the doing of it rather than in the end of it.[16]

With Mandela's release on 11 February 1990, the Picket had to come to an end. By keeping going for another 13 days after his release, City Group created space to celebrate their own achievements and to walk away from the site of the Non-Stop Picket in the manner of their choosing.

To mark the end of the Picket, a special issue of the Group's newsletter, *Non-Stop Against Apartheid*, was produced.[17] On its front page, surrounding the now iconic photo of Nelson Mandela walking out of Victor Verster Prison, hand in hand with Winnie Mandela, both of them giving clenched fist salutes, the headline was a call to further action – "Mandela is free ... now smash apartheid."

On the back cover, City Group offered a more thorough assessment of the situation in South Africa and renewed their pledge to continue campaigning:

On 11 February 1990, Nelson Mandela was released from prison. Other veteran prisoners had been released in the previous year, including Walter Sisulu, Govan Mbeki, Ahmed Kathrada of the ANC and Jafta Masemola and President Zephaniah Mothopeng of the PAC. On 2 February 1990 the ANC, PAC, AZAPO and SACP were unbanned. City AA welcomes all these reforms, won by the freedom struggle led by the black majority. Nevertheless we know that there is a long way to go in the struggle. All the main pillars of apartheid remain in place, and majority rule is yet to be won. Thatcher has quickly moved to lift sanctions against the regime.

We have remained on the Non-Stop Picket, as we pledged to, until Nelson Mandela is free. Now that the Non-Stop Picket is ending we have pledged ourselves to continue and to escalate our solidarity.[18]

The article went on to list City Group's planned actions for the coming months. Part of the intention of the Non-Stop Picket 'victory' rally was to keep supporters engaged in campaigning against apartheid. Although the Picket's focus on calling for Mandela's release resonated with a wide layer of people and made it (relatively) easy to engage them in anti-apartheid activism, it may also have made it harder to sustain large numbers of activists in solidarity work during the remaining years of the transition to a post-apartheid state. City Group continued to campaign against apartheid until 1994, but its membership and the size of the core activist group tailed off in the years after the Picket ended.

The rest of the commemorative issue of *Non-Stop Against Apartheid* mostly contains photos of key events from the four years of the Non-Stop Picket, along with short, year by year commentaries on City Group's campaigning. Many of the events noted in the issue have been described in this book (for example, the Hands Off Women Picketers campaign from 1986, the first anniversary of the Picket, the campaign to defeat the police ban on protests in front of the Embassy in 1987, and the 1,000 days and nights celebration, amongst many others). This commemorative issue gives a good indication of the events City Group activists viewed as significant at the time.

Over the four years of its existence, the Non-Stop Picket had mobilised countless young people to active campaigning against apartheid. Some of them were deeply committed to the anti-apartheid cause before they joined the Non-Stop Picket; others were angry or lonely and came to understand the issues involved in apartheid through the companionship they found on the picket. Whatever else was going on in their lives, for many of the young (and not so young) activists who sustained it, the Picket was somewhere they could feel they belonged. The prospect of that sense of belonging ending with the end of the Non-Stop Picket was deeply unsettling.

Given how important the Non-Stop Picket was in their lives, we were startled that so few of the former picketers who we interviewed (and who knew they were present on that day) had very clear memories of the Picket's final celebration. Nicole, for example, had the following to say:

> Yes I was [there], but I don't really recall it apart from afterwards where I have recollections of being on the tube singing songs ... I've no idea where we were heading.[19]

Similarly, Francis recalled:

> My memory isn't totally clear. We had a march, which included walking down Whitehall. There was a party afterwards, (I can't remember where) but I do remember that Comrade Virman Man was on particularly impressive form as he led City Group Singers with his fine strong bass voice. Comrade Rene turned to me and said "no one can stop City Group"![20]

We would suggest that these hazy memories are precisely a measure of the confusion and contradictory emotions many participants experienced that day. In contrast to the clear memories they hold of other events surrounding the Non-Stop Picket, that final rally was disorienting (Maddrell 2016). While City Group members wanted to celebrate their achievements, many experienced a sense of anti-climax that the Picket was finally over. Having been 'non-stop' for nearly four years, many picketers found themselves contemplating what it would mean to lead their lives at a different pace and without the comforting knowledge that they could always find companionship on a pavement in Trafalgar Square. Andre Schott, for example, remembers:

> It was a bit strange, it just didn't feel real. I remember the decision was made and I agreed with it, because it wouldn't be sustainable beyond Mandela's release. It was just a bit, yeah, you didn't want to let go in a way. But apart from that I can't remember, I remember marching away with lots of flags and banners.[21]

Similarly, Deirdre Healy recalled that the sense of victory and success was overshadowed by a fear of what would happen to her and her friends without the Picket to cohere them:

> I suppose that would have been a side thought definitely, but there was some of that. And that oh my god, it was going to finish, as a political action it was going to end. And that we had felt such belonging to and were so invested in, we were all hugely invested in it. Well I can only talk for myself, there was that huge level of personal investment in this thing, and oh my god Mandela was going to be released.[22]

For Richard Roques, the Group's Treasurer, who had been involved with City Group since its origins in the Free Steve Kitson campaign in early 1982, the end of the Non-Stop Picket was extra-ordinarily emotional:

> I burst into tears ... Well I knew it had to end and I wanted it to end, but I also knew that it was sort of an end of something that was very

important to me, and that in a sense we knew that we should continue. Because the release of Nelson Mandela wasn't what it was all about, and actually we were right about that. So yeah, I found it very moving.[23]

Andrew Privett was even more direct and acknowledged that that day and the period that followed it was marked by "a sense of bereavement" for the loss of the Non-Stop Picket and everything that accompanied and surrounded it.[24] He talked about feeling a "lack of focus" that was "partly due to that sense of bereavement, partly burnout." He was not the only picketer to talk of how tired and fatigued they felt by the time the Non-Stop Picket ended. For this reason, many of the most committed members of the Group accepted that, however important time spent on the picket and with fellow picketers was for their lives, it was time for them to stop trying to maintain a continuous protest at the Embassy. In his interview, Simon Murray was very blunt about how difficult the Picket had been to keep going over its final winter:

> Towards the end there was, well, I think all the way through the Picket there was a main group of people who kept it going, but towards the end I think it had got down to about perhaps 20, 30 people trying desperately hard to keep it as 24 hour as it possibly could be and it sometimes did feel like hard work. A lot of people had drifted away, a lot of good people and it was hard towards the end and I think yeah, to be honest with you, I think that it was a big relief.[25]

If many core picketers felt a sense of relief at the Picket ending (as well as a certain amount of guilt for feeling like that about it), the final rally was more subdued than many of the larger rallies that City Group was used to organising as a focus for its campaigns (see Chapter 4). Even so, the picketers left with a sense of success and defiance. During the final rally, graffiti was scrawled on the pavement beneath the Picket by picketers who were screened from police surveillance by the weight of the crowd that afternoon. These scribbled texts, written in black indelible ink, link the Picket's political cause closely to its location. The first proclaimed: "We won the fight, Amandla!" quickly followed by the promise: "If your [sic] lying we'll be back!!" The South African slogan, *Amandla!* (power) and the threat that the Picket would return, if necessary, demonstrates the stretched spatiality of the Non-Stop Picket, linking the anti-apartheid struggle in South Africa with solidarity activists in the heart of London (Katz 2001; Massey 2008). The second piece of graffiti celebrates the duration of the Picket – "City AA '86–90" – but stakes a territorial claim over that stretch of pavement (Leitner et al. 2008; Halvorsen 2015), indeed the whole of Trafalgar Square – "We won the Square." Over the four years of the Non-Stop Picket, that stretch of pavement became more than just the location of a solidarity protest related to a national liberation struggle thousands of miles away; it became a home of sorts to the eclectic mix of (mostly young) people who joined the protest (Brown et al. 2017). However, as much as they celebrated their success and the longevity of their protest, they were reluctant to completely let go of the small territory they had made their own.[26]

Continuing solidarity

Following the end of the Non-Stop Picket, City Group continued its anti-apartheid campaigning. Throughout 1990, they led high-profile solidarity campaigns for the Upington 14 on trial in South Africa. In August 1990, 11 City Group activists were arrested during a consumer boycott action at a supermarket in South London where they filled shopping trolleys with South African goods, took them to the checkouts, and loudly refused to pay for them.[27] On this occasion, their protest was part of an on-going campaign of solidarity with striking fruit farmers from Zebediela in South Africa (which also involved City Group activists and a leading member of the Black Consciousness organisation, AZAPO, occupying the offices of *The Guardian* newspaper in protest at their refusal to report the strike). Alongside these specific campaigns, they kept going with 36-hour pickets of the South African Embassy every weekend. Although many members did drift away after the end of the Non-Stop Picket, these weekend pickets were relatively easy to sustain throughout 1990; but they became progressively harder to keep going, with the Group relying on the energies of a shrinking pool of active supporters. As Mark Farmaner, who remained centrally involved in the Group until 1994, confessed:

> After Mandela was released I wanted it to stop. Sometimes I changed my mind but generally I think it had to, and it just dragged on with that weekend picket and I know people were saying the weekend picket had to happen because that's where all the funds came from and if we didn't have the weekend picket we wouldn't have the money to carry on doing other stuff. But I think it just, it's always that thing with NGOs and, you know, they never want to stop doing what they're doing or shut themselves down ... So I think it would have been better just to have stopped right there than sort of struggle on with that. I'm not sure what we really achieved after the Mandela release and the weekend picket. I'm not sure that there were anything, successes you could point to by carrying on. We were still raising money but I don't know what we were doing with it.[28]

It is not our intention to offer an evaluative assessment of how effective the Group was in the years after the Non-Stop Picket ended. However, in examining how picketers responded to the end of that continuous protest and assessing the impact that the Non-Stop Picket had on the subsequent lives of its (young) participants, it is useful to consider what happened next.

In that context, we examine two sets of events from the years after the Picket ended. First, we examine how members of City Group attempted to gain retrospective recognition for their protest from Nelson Mandela and other freed ANC leaders. But, mainly, we examine the consequences of a series of arrests that took place the night before a planned protest at a rugby match to be played by the South African Springbok team in Leicester, which the Group believed was in contravention of the international sports boycott (despite the rapidly changing political situation in South Africa at the time).

When it was announced that Mandela would be visiting London in April 1990 (for the first time since his release), several members of City Group wrote to him requesting that he made time during his visit to meet the people who had maintained a four-year non-stop picket for his release. The meeting never happened (and, as far as we know, many of the letters received no acknowledgement or reply). These letters were part of a political campaign to gain Mandela's (retrospective) approval for the Non-Stop Picket in the face of long-standing opposition to it from sections of the London ANC (Kitson 1987; Trewhela 2009; see also Chapter 2).

When Mandela spoke at the Wembley concert in April 1990, those City Group members who were in attendance, and others who watched it on television, took solace from the opening lines of his speech:

> Our first simple and happy task is to say thank you. Thank you very much to you all. Thank you that you chose to care, because you could have decided otherwise. Thank you that you elected not to forget, because our fate could have been a passing concern. We are here today because for almost three decades you sustained a campaign for the unconditional release of all South African political prisoners. We are here because you took the humane decision that you could not ignore the inhumanity represented by the apartheid system.[29]

Having stood outside the South African Embassy calling for the unconditional release of Nelson Mandela for so long, City Group activists took comfort that they were (probably) amongst those being thanked for their solidarity action, for caring. His courage and leadership (but not necessarily his alone) had inspired them to take militant direct action against apartheid, even as they understood him as but one man in a popular mass movement.

Although filled with admiration and respect, the letters that City Group activists sent to Mandela in the months following his release were more than just political 'fan' letters; they were a political act. City Group members knew that if they could persuade Mandela to come and speak to them, it would be a major blow against all those in the London ANC and the British Anti-Apartheid Movement who had sidelined and rubbished City Group and the Kitson family over the previous eight years (see Chapter 2). They felt that City Group's case for 'non-sectarian solidarity' with all the liberation movements, and a militant, street-based solidarity movement would be vindicated. Of course, whatever Mandela himself thought of the Non-Stop

Picket and City Group's politics, his close advisers were never going to let that meeting happen – it would have gone against the plans for a negotiated, diplomatic settlement to end apartheid (Bond 2000; Gibson 2011, 2012).

All the time that those multi-party negotiations were taking place, City Group activists continued to organise militant direct action against the representatives of apartheid. When it was announced that the South African Springboks rugby team was due to tour Britain during the autumn of 1992, City Group sprang into action. Specifically, they responded to the call from South African anti-apartheid organisations for the tour to be challenged and opposed. Despite the on-going multi-party talks and negotiations in South Africa that were attempting to manage the transition to a post-apartheid state, the non-racial South African Council on Sport (SACOS) asserted that apartheid was not over, and the sports boycott should stay in place.

Support for City Group's stance came from various quarters. Mpotseng Jairus Kgokong, the Secretary General of the Black Consciousness Movement (of Azania) (BCMA) faxed City Group from Harare on 29 September stating:

> The BCM(A) is pleased to note that not everybody is duped by the tricks of De Klerk. It is in this light to the BCM(A) takes this opportunity to fully support your campaign against the coming Springbok tour in November this year. The BCM(A)'s firm standpoint coincides with yours: there should be no sporting contact with South Africa until there is democratically elected Constituent Assembly in place.[30]

Along with British-based members of the BCM(A) and the PAC, City Group helped convene the Springbok Reception Committee to oppose the tour. An undated fax from SACOS to City Group offered their "heartfelt gratitude for your support."[31] Between August and early November, City Group assembled international support for the Springbok Reception Committee from the Azania Komitee of the Netherlands, the Azania Committee in Sweden, and South African organisations including the National and Olympic Sports Congress of South Africa, the Amateur Swimming Association of South Africa, and the South African Amateur Athletics Board. On 6 October, Don Nkadimeng, the Secretary General of AZAPO (the main Black Consciousness organisation inside South Africa) added his support with this statement:

> AZAPO is fully behind your efforts to promote our slogan "No normal sport in an abnormal society", and we will be watching events from here with keen interests. I can confidently tell you that over 40 million Blacks wish you success in your efforts.[32]

Over the summer, ahead of the tour, City Group circulated letters to local anti-apartheid groups, trade union branches and student unions calling for demonstrations to greet the Springbok team on their arrival in the UK and in every location where they were due to play (Bath, Bristol, Leeds, Leicester, and Twickenham). Of those cities outside London where the Springboks had matches scheduled, the Springbok Reception Committee received the most positive support from a group of activists in Leicester, which included Ross Galbraith and Gary Sheriff at its core (City Group had previously worked closely with them after they were sacked for refusing to work on a contract for South Africa at the factory where they worked).[33]

Throughout this period, the public announcements from the British Anti-Apartheid Movement were muted and (potentially) contradictory. In large part, this appears to be because they did not want to be seen to publicly contradict the ANC who had appeared to suggest that the sports boycott was no longer necessary, as the South African government had begun repealing apartheid legislation. In 1992, the ANC position appeared to be that they were not opposed to international sporting fixtures against South African teams as long as they were played non-racially. Fieldhouse (2005: 455) argues that, in private, leading Anti-Apartheid Movement (AAM) members questioned the logic of this position with the ANC leadership. On 9 October, the AAM issued a confusing statement to its local groups, seemingly suggesting that they were no longer participating in a boycott of sporting links with South Africa but would protest President De Klerk's presence at the match he was due to attend. However, later that month, the ANC appeared to rethink its position yet again. On 27 October, Richard Roques, from the Springbok Reception Committee, wrote to the London ANC in following terms:

> We welcome the announcement by the ANC to discontinue their support for the forthcoming tour of the Springbok Rugby team to Britain. Following a unanimous decision at our AGM last year the City of London Anti-Apartheid Group wrote to the ANC to inform them that we felt the decision to suspend the sports boycott was premature. Earlier this year we ran on the track when Zola Budd ran (and dropped out) of a race at Crystal Palace. We ran on the pitch and pulled up the stumps when the Transvaal cricket team played at Lords. When we learnt of the forthcoming Springbok tour we formed the Springbok Reception Committee with the support of the [PAC], AZAPO and the South African Council on Sport. The response of the British Anti-Apartheid Movement was to phone the local Leicester Group who are active members of the Committee to tell them that the sports boycott was a 'grey area'. We welcome their renewed commitment to the sports boycott but deplore their attempts to discourage what action was taking place before it received their 'official' seal of approval...[34]

What is striking about this letter is the (uncharacteristic) lack of deference offered to the ANC. In speaking on behalf of City Group, Richard Roques clearly felt confident enough to tell them how they should pursue anti-apartheid campaigning internationally.

Even without the active support of the ANC, the Springbok Reception Committee was a dynamic and militant campaign. Francis described the campaign like this:

> We hounded the players wherever they went, not only holding rowdy demos, but letting off stink bombs, egging and flouring them etc.[35]

On 5 November 1992, *The Independent* and other papers reported that nine anti-apartheid activists had been arrested in the vicinity of the Leicester Tigers' rugby stadium.[36] The nine, who became known as the Springbok 9, were City Group members and supporters. They were arrested by the police as they approached Central Leicester in a white van with a ladder attached to the back, containing items that the police believed were to be used to cause criminal damage to the rugby pitch. Several of the arrestees and other members of their defence campaign believe that the police were tipped off by an informer inside the campaign (Evans and Lewis 2013). When they were committed for trial, in March 1993, the *Leicester Mercury* (24 March) reported:

> Members of City of London AA group – including its secretary and a photographer... were in a van which contained bags of broken glass, metal tacks, balaclavas, gloves and two notices which read "Danger, do not play on the pitch."[37]

As Ross Galbraith explained, the local group in Leicester was expecting a delegation from City Group and had arranged accommodation for them in the city at the Socialist Centre, but had not enquired too closely as to what their exact plans were.[38] Their first inkling of the arrests came when it became apparent that the Group had not arrived to use their city-centre crash accommodation. The Leicester group became central to the work of the Springbok 9 Defence Campaign, supporting them through their initial court hearings in Leicester and their eventual trial at Nottingham Crown Court. As two of the nine recounted:

> We were arrested and charged with "being in possession of articles likely to cause damage to the Leicester rugby football ground". It sounds innocuous but carried maximum penalty of 10 years in prison. Quite an experience!... [There was a] two-week crown court trial... We argued a political case. I remember arriving in court on the last day carrying a little bag with my toothbrush in, fully expecting to get sent to prison – but there was a hung jury and the judge bound us over to keep peace for two years.[39]

> We had a two week court case, we managed to get it moved from Leicester, because we said local opinion would be against us and we had a court case in Nottingham on conspiracy to cause £250,000 worth of

damage to the Tigers' pitch, which are quite serious charges. And we could have gone down on, but we had a very good defence team: Courtenay Griffiths, Vera Baird, Ken Macdonald and John Riley, who put up a great defence and a jury that was just about hung with two people on it not prepared to convict. But anyway, yes, so we thought it was important to target South African sport.[40]

With the prospect of nine of their long-term activists being imprisoned, for lengthy sentences, the Springbok 9 Defence Campaign became one of the central preoccupations of City Group's campaigning during the last year of its existence. What began as an act of solidarity with the black majority in South Africa during the final period of apartheid ended as a campaign-building solidarity for militant British anti-apartheid activists. At a time when the ANC was prepared to make concessions to the negotiations process in South Africa, City Group took its lead from those sections of the liberation movement in South Africa that were more suspicious about what the negotiations would deliver. Throughout its existence, City Group encouraged and supported its members to take direct action and break the law as a tactic for protesting Britain's links with apartheid South Africa. Many hundreds of arrests were made during the course of the Group's 12 years of anti-apartheid campaigning, and a small number of activists spent time in prison as a result of their actions. However, many of the charges that followed the majority of these arrests were for relatively minor public order offences. The case of the Springbok 9 suggests the possibility that, as the core of the Group shrunk, and they became even further marginalized from mainstream anti-apartheid campaigning, some in the Group may have been prepared to take higher risks to draw attention to their cause.

Making sense of disappointing victories

> On the positive side, this is the only campaign I have been involved in that actually achieved its goals! The wars I protested against happened anyway, with consequences even worse than I feared, the mines closed down... and I could go on.[41]

When asked to reflect on post-apartheid South Africa, many former anti-apartheid campaigners associated with the Non-Stop Picket paused before answering. They often paused a little too long, as if they were struggling to find the right words or trying to work out how they could (or should) respond. A few former picketers, mostly those whose opposition to apartheid had been motivated by a religiously inspired liberal humanitarianism celebrated the end of apartheid as an unambiguously positive outcome for human rights in South Africa (and, to some extent, globally). However, the majority of former solidarity activists were (after some hesitation) more balanced, if not explicitly critical of the post-apartheid settlement and the impact of two decades of ANC rule. The more 'radical' that picketers were in how they envisaged apartheid might end, the more they expressed disappointment with the form that the post-apartheid transition took.

A number of features of contemporary South African politics and society were identified over and over again by former solidarity activists as evidence of the failings of the post-apartheid settlement (and to explain their disappointment with the ANC in power). They acknowledged that there was still (excessive) economic inequality within South African society, with extremes of wealth and a majority black underclass still living in dire poverty. The continuing existence of improvised, informal settlements after two decades of ANC rule worried several interviewees. They also critiqued the ANC's handling of the AIDS crisis in South Africa (Schneider 2002; Tucker 2016). Several also targeted the ANC, and specific members of its leadership, for its involvement in a range of corruption scandals (Feinstein 2009). Furthermore, several interviews were conducted shortly after the Marikana massacre of striking miners in August 2012 (Alexander 2013), and this both angered and disillusioned many. Daniel linked the track record of the ANC in power back to some of the criticisms levelled at City Group by others in the Anti-Apartheid Movement in the early 1990s:

> Well it's a mess isn't it? Yeah the AAM said we [didn't] know when to stop, but then the thing was that the ANC kind of stopped more or less immediately they got into power.[42]

Criticizing the ANC did not come easily to City Group members, despite the tensions between the Group, the Kitson family, and leading members of the ANC in London throughout the 1980s. Even though many City Group members ended that decade being more sympathetic to the politics of the PAC and AZAPO (or 'workerist' tendencies within the South African trade union movement, such as Moses Mayekiso) than the ANC, they remained ethically committed to providing 'non-sectarian solidarity' to all wings of the South African liberation movement(s) (see Chapters 1 and 2). Their solidarity was unconditional, and members of the Group were practiced in not publicly criticizing actions and decisions of the liberation movements that they might have found questionable in private. Even after the ANC had been in government for 20 years, criticizing their policies and conduct did not come easily for some former solidarity activists. For example, the question that Andy Privett threw back at us when we asked him about the post-apartheid settlement was very telling:

> Do I have to be diplomatic? Disappointed. It is not my place to dictate to the black majority, but the negotiated settlement was an economic settlement in the interests of international big business. It was a failed revolution – people who should be benefitting from the wealth of the

country aren't. But, in the same circumstances, would go back and do same things again – an evil is an evil. Deeply disappointed.[43]

Andy quickly qualified his 'disappointment' by acknowledging that "it is not my place to dictate to the black majority."[44] This was the dilemma faced by many supporters of the Non-Stop Picket. Some had joined the Picket as convinced communists and anti-imperialists; whilst many others were won to an anti-capitalist perspective through their involvement. As solidarity activists, they saw their role as being to support and enable the majority of the South African people to reach a position where they could determine their own political future. Many believed, faithfully, that apartheid would not be ended without a revolutionary upheaval. The township uprisings of the mid-1980s, the upsurge of trade union militancy, and the volume of protesters that could be mobilised by the United Democratic Front and, later, the Mass Democratic Movement helped convince them of the veracity of this belief. Indeed, one interviewee, April, argued that the nature of the negotiated settlement took the form it did precisely to prevent the advancement of 'communist' practices in the South African townships and trade unions (Bell 2009):

> I think it's really sad. But I think we knew that. I think that when they freed Mandela, they had to do it because of what was happening on the ground in South Africa. And it was so exciting, wasn't it? With the street committees and area committees – people in the townships were organising along properly organised, well, communist, ways of doing things. And they had to stop it.[45]

Georgina was one of those who acknowledged her expectation that post-apartheid South Africa would be socialist:

> I did think that it would probably be – I suppose I took it for granted there'd be a stronger socialist line and obviously that's a disappointment.[46]

Similarly, Susan articulated the belief that the struggle against apartheid was an anti-capitalist struggle (and exemplified the sense of disappointment that a post-capitalist transformation of South African society was not the outcome of how apartheid ended):

> I was very disappointed really because well I suppose when we were struggling for Nelson Mandela to be freed we wanted to see equality in Africa, and because that struggle also involved the struggle against capitalism in many ways, in terms of the exploitation of the African people, I suppose we wanted to see them start to build a society based on the people and not profit based on education, health care, etc., socialist society, and that wasn't the case.[47]

Some former picketers were even more brutally honest in their assessment of the post-apartheid settlement. For Richard Roques:

> It's a disaster. And we've got a black bourgeoisie who've nearly replaced the white oppressors, and what you've got is capitalism in South Africa, and though we weren't dictating to the liberation movement, we were supporting a revolution, we were supporting people who were going to change the conditions they lived in. We weren't supporting people just to get the vote, so they've won the vote and not won their freedom, and I think Nelson Mandela is a sell-out.[48]

> Well I am glad that now all South Africans have a chance to determine their future, more of a chance then they did then. I am a communist, so I would have liked the outcome to be a revolution that represented the interests of the black majority and that hasn't happened. That is something for the future. So I see what happened in City Group, and what it did, as belonging to a particular period of history. Now it's moved on from that. The revolution has been stalled effectively by the deal that was done at the point when Mandela was released and the first elections were held.[49]

This highlights that members of City Group held a range of political perspectives. Even amongst those advocating a revolutionary end to apartheid, there were a range of perspectives. A minority of picketers, at the time, held more anarchistic or Left communist views and were sceptical that the ANC (or the PAC) would deliver an anti-capitalist social transformation, no matter how much they publicly articulated socialist rhetoric. Nick Manley, was more sanguine, remarking:

> To me, it's gone more or less as I expected. But I do sometimes wonder how people feel who were convinced that the ANC would bring in a socialist utopia for everybody.[50]

Of course, with a distance of more than a quarter of a century since Nelson Mandela's release, and two decades since the end of apartheid, former picketers' reflections on the post-apartheid settlement are coloured by the ways in which their political beliefs have developed and mutated in the intervening period.

When asked what she thought about how post-apartheid South Africa turned out, Simone Maloney answered with one simple word – "capitalism."[51] While this represented a specific 'disappointment' for many former picketers, others recognised that the transition to 'non-racial' capitalism in South Africa had removed the 'abnormal' aspects of apartheid society and clarified the need for on-going anti-capitalist organising and working-class politics there. In different ways, both Ross Galbraith and Lorna Reid articulated this position:

> I suppose I think the first point about South Africa is that it's kind of like Ireland in that it's now just 'normal politics'. You've taken the apartheid bit out of it and it's now just a class struggle thing that's going down there. Colour doesn't come into it anymore. No! *It doesn't*. It's a pure class issue now there and they're going to have to go through all that and that's going to have to develop. But I suppose, in a way, you could say that we... that we *helped*. ... Everybody who was involved outside South Africa. It's helped to take the apartheid bit out of it and they can now carry on with their own struggle.[52]

> Yeah, so to expect something to develop within twenty years is not going to happen. It may never happen, but that is, now that is the choice of the South African people. You know, apartheid had to be removed in order to allow people their choices, but working class South Africans will need to become organised. Just the same as the working class here need to become organised, and they're not. So, you know, where do we turn our attentions? You didn't quite do what we expected you to do, all that time in the rain outside the Embassy, this is the thanks that we get, you what I mean, it's a crazy position to take, isn't it?[53]

Despite their disappointment, many former solidarity activists took solace from the fact that just societies cannot be achieved quickly. In many ways, their acknowledgment that lasting and significant social change is a long-term process seems at odds with the urgency and pace of their erstwhile 'non-stop' protest to end apartheid (Sharma 2014). Now, rather than talk of rupture and revolution, their temporal understandings of social change are articulated more in terms of emergence and development. As Helen Marsden recognised:

> I think it's turned out the way that any kind of emerging economy which is so heavily dependent on the big guns would have turned out, sadly. But at the same time, I don't think that that means that the work that we or anybody else who was involved in anti-apartheid work, you know, I don't think that that means that our work wasn't important.[54]

In accepting that their previous goals of an economically (as well as a racially) just and equal South Africa might take much longer to achieve than they had hoped, picketers (like Helen) were careful not to suggest that their solidarity and their protests had been for nothing. Indeed, some remembered that more experienced South African revolutionaries had always warned them that the 'real' struggle would begin once apartheid had ended. For example, Simon Murray recalled words of warning, guidance, and encouragement offered by David Kitson in the early 1990s:

> I think one of the most poignant things I ever heard from anybody was when David Kitson was over the last time when the Picket was still

going on and he said, don't think when apartheid ends that that's when your job ends, he said, it'll probably be more important for you to be around after apartheid's finished.⁵⁵

While it is instructive to analyse how former members of the Non-Stop Picket reflect on post-apartheid South Africa and have dealt with the 'disappointments' attached to the victory of the struggle against apartheid that they had supported so vigorously, what is more interesting is to consider how they have utilized the lessons of their picketing (and that disappointment) in other contexts. In the concluding chapter, then, we turn our attention to picketers' lives since the end of the Non-Stop Picket. In doing so, we consider how they think the experience of the Picket has influenced their subsequent lives. We consider their personal and professional lives, as well as the extent to which they have (or have not) remained active in political campaigning. We consider how they may have continued to utilise and rework knowledge, skills, and practices that they acquired or developed during their time being non-stop against apartheid. Finally, we conclude by reflecting on some of the lessons that the Non-Stop Picket might offer for those interested in practising transnational solidarity today.

Notes

1 Interview with Andre Schott.
2 Interview with Carol Brickley.
3 City Group (1989), *Annual Report 1988–89: 365 Days in Struggle* (23 September).
4 C. Brickley (1989), 'Apartheid and imperialism', *Non-Stop Against Apartheid*, 36 (October/November), p. 5.
5 Letter from S. Murray to City Group Committee, 21 August 1989, stating that he appreciates the work and guidance of the RCG but does not want to feel pressured into joining them. Gavin was also one of the 'dissenters' at this time.
6 C. Brickley (1989), 'Apartheid and imperialism', *Non-Stop Against Apartheid*, 36 (October/November), p. 5.
7 A. Schott (1989), 'Stop the hangings. Free the Upington 26', , 34 (June), p. 1, 3; City Group (1989), 'Common purpose. The Upington 14 must not hang', , 35 (July/August), p. 1; City Group (1990), 'The Upington 14 must not hang', *Non-Stop Against Apartheid*, Special issue (February), pp. 1–8.
8 City Group (1989), 'Veteran ANC and PAC freedom fighters released', *Non-Stop Against Apartheid*, 36 (October/November), p. 1.
9 Letter from City Group to members and supporters titled 'Forthcoming release of Nelson Mandela', 18 January 1990.
10 J. Shields (1990), 'Nelson looks on as London celebrates', *The Daily Telegraph*, 12 February. A. Hamilton (1990), 'Trafalgar Square delight', *The Times*, 12 February. The ITN Archive contains raw footage reporting from outside the Embassy on that day, including interviews with Carol Brickley and Andre Schott about the Non-Stop Picket: www.itnsource.com/en/shotlist/ITN/1990/02/11/BSP110290007 (last accessed 28 August 2016).
11 In 1968, as the Minister of Technology in the 1964–70 Labour Government, Tony Benn signed a contract with Rio Tinto Zinc to extract 7,500 tonnes of Uranium from the Rossing mine in Namibia in contravention of UN sanctions (due to

194 *'Until Mandela is free ...'*

South Africa's illegal occupation of Namibia) (Roberts 1980). See also: Hansard HL Deb vol. 364 cols. 1226–34, 20 October 1975.
12 Interview with Francis Squire.
13 Interview with Jacky Sutton.
14 Interview with Nicole Aebi.
15 Interview with Betta Garavaldi.
16 Interview with Carol Brickley.
17 City Group (1990), 'Mandela is Free. ANC, PAC and AZAPO unbanned. Now smash apartheid!', *Non-Stop Against Apartheid*, 38 (February).
18 City Group (1990), 'Mandela Free – Smash Apartheid', *Non-Stop Against Apartheid*, 38 (February), p. 12.
19 Interview with Nicole Aebi.
20 Interview with Francis Squire.
21 Interview with Andre Schott.
22 Interview with Deirdre Healy
23 Interview with Richard Roques.
24 Interview with Andrew Privett.
25 Interview with Simon Murray.
26 See Halvorsen (2017) for a discussion of similar tensions at the 2011 Occupy London protest camp; and Russell et al. (2017) for a discussion of the problematic refrain of camping in British Climate Camps.
27 Press statement about the arrest of 11 City of London Anti-Apartheid Group protestors against Safeway's, in Walworth Road, stocking South African fruit and vegetables – in solidarity with Zebediela strikers, 17 August 1990.
28 Interview with Mark Farmaner.
29 N. Mandela speech at Wembley Stadium, London, 16 April 1990. Full transcript available at: http://db.nelsonmandela.org/speeches/pub_view.asp?pg=item&ItemID=NMS027&txtstr=Wembley (last accessed 28 August 2016).
30 Faxed letter from Mpotseng Jairus Kgokong, Secretary General of the Black Consciousness Movement (of Azania), 29 September 1992.
31 Faxed letter from SACOS Rugby Union to City Group, 1992 (undated).
32 Faxed letter from Don Nkadimeng, Secretary General of AZAPO, to City Group, 6 October 1992.
33 Interview with Ross Galbraith.
34 Letter from R. Roques, on behalf of the Springbok Reception Committee, to the London ANC, 27 October 1992.
35 Interview with Francis Squire.
36 'Police arrest nine protesters', *The Independent*, 5 November 1992, p. 45.
37 Leicester Mercury (1993), 'Anti-apartheid demonstrators face rugby match charges. Nine committed for trial', *Leicester Mercury*, 24 March, p. 18.
38 Interview with Ross Galbraith.
39 Interview with Andrew Privett. The black South African photographer arrested with the members of the Springboks Reception Committee was found not guilty at this trial.
40 Interview with Cat Wiener.
41 Interview with Helen Landau.
42 Interview with Daniel Jewesbury.
43 Interview with Andrew Privett.
44 Ibid.
45 Interview with April.
46 Interview with Georgina Lansbury.
47 Interview with Susan Yaffe.
48 Interview with Richard Roques.
49 Interview with Carol Brickley.

50 Interview with Nick Manley.
51 Interview with Simone Maloney.
52 Interview with Ross Galbraith.
53 Interview with Lorna Reid.
54 Interview with Helen Marsden.
55 Interview with Simon Murray.

9 Lessons and reflections

On 14 March 1987, 5,000 people marched six miles across London on City Group's 'March for Mandela'.[1] The demonstration started at Whittington Park in North London and ended at the South African Embassy in Trafalgar Square, carried by the rhythms of the Batucada Mandela samba troupe. It was one of the largest protests that City Group ever organised.

The participants in the March for Mandela were described by Carol Brickley, City Group's Convenor, in her speech on that day as "Thatcher's rejects."[2] In many ways, they were. The demonstration included contingents of miners from Hatfield Main NUM branch, from the Viraj Mendis Defence Campaign, and other anti-deportation groups, lesbian and gay activists from the Wombourne 12 Defence Campaign, and students fresh from the anti-apartheid occupation at the London School of Economics (LSE). The demonstration represented a temporary coalition of sections of British society that had been marginalized, repressed, and/or radicalized by eight years of Thatcherism. The March for Mandela provides a snapshot of British radicalism in the 1980s and insights into the flows of intersectional solidarity that circulated at the time (Kelliher 2014, 2017).[3]

This concluding chapter examines the impact that participating in the Non-Stop Picket has had on the personal and political lives of former picketers who may have understood themselves as amongst the 'rejects' of Thatcherism. Some were never politically active following the Non-Stop Picket; others have never stopped their political activism. Either way, most believe that their time on the Non-Stop Picket profoundly shaped their lives. Many former activists can identify specific skills and competencies they developed through their involvement in the Non-Stop Picket that they continue to draw on as resources in their adult lives (now even the pre-teens involved in the Picket are reaching their forties). In discussing the afterlife of the Non-Stop Picket, this chapter examines how the space of the Non-Stop Picket and the practices through which it operated not only created a high-profile expression of solidarity with those resisting apartheid in Southern Africa, but also fostered relationships of care and comradeship that helped participants deal with the other pressures affecting their lives (beyond the pressures of campaigning non-stop).

The book concludes with lessons from the Non-Stop Picket for those interested in urban social movements (and protest camps), young people's activism, and the history of the international movement against apartheid. It advocates a social practices approach to activism that offers new possibilities for understanding the dynamic ways in which activist practices become bundled with other aspects of life and life-course transitions. In doing so, it extends the reach of recent discussions about the transformative effects of practising solidarity (Featherstone 2012; Kelliher 2017). As we have argued, young activists' political commitments are entangled with the everyday politics of growing up, and youthful activism is not simply a 'phase' to be outgrown; it can be a positive resource for socially engaged adulthood.

Comradeship and care

In today's policy parlance, many of the young people who participated in the Non-Stop Picket were (or, through their participation, made themselves) 'vulnerable' – if only through their brushes with the criminal justice system (Pain 2004; Pain and Francis 2004; Gaskell 2008). A number of picketers remembered moments when they (or others) became 'vulnerable' through their involvement with anti-apartheid protests. Sometimes these reflections came from people who were young picketers at the time. They also came from picketers who now work with young people and reflected on those moments of vulnerability (including questioning whether City Group always did enough to protect its young protagonists). Without presenting political protest as problematic for making young people 'vulnerable', we acknowledge some of the (unresolved) tensions around this vulnerability. The intensity and duration of the picketers' collective commitment to maintaining a 'non-stop' protest could exaggerate their vulnerability. At the same time, the relations of comradely care fostered on the Non-Stop Picket (often) helped participants to cope with the personal problems that could make them 'vulnerable' in various ways (Butler et al. 2016).

Among the young people who joined the Non-Stop Picket, some were already 'vulnerable', and not just the young homeless people who spent time there. Through their involvement with the Non-Stop Picket, many other young people (from relatively safe and stable home environments) potentially made themselves 'vulnerable'. They stood on a pavement in Central London voicing their support for armed struggle against oppression in a foreign country with strong links to their own government. Although not universally supported within the Group, the Picket often displayed the iconography of the armed struggle and the insurrectionary violence of the township uprisings: They marched with cardboard cut-outs of AK-47s or 'necklaces',[4] sang the songs of a guerrilla army, and chanted slogans advocating political violence – the Pan-Africanist Congress (PAC) slogan 'One Settler, one bullet!' had a certain taboo appeal.

As well as having a 'radical', anti-imperialist perspective on the struggle against apartheid, many young picketers also developed wider anti-capitalist and anti-authoritarian perspectives through their discussions, debates, and experiences on the Non-Stop Picket. The experience of violence, abuse, or injustice at the hands of the Metropolitan Police often triggered a radicalisation of picketers' analysis and action. As discussed in Chapter 6, City Group recognised that certain groups of picketers were personally or politically subject to disproportionate targeting by the Police. They responded politically to that threat. Young black men were repeatedly arrested, some women picketers were sexually assaulted whilst in police custody, and picketers with mental health issues or learning disabilities often required a particular kind of support to ensure neither they nor the police escalated small conflicts unnecessarily. Ken, a blind black picketer explained:

> The police were particularly hostile to young black people. And they hated the fact of young black people coming into any sort of organised politics. It was all well and good being able to pick on individuals and lock up individuals, but seeing a group of young black people on the picket or associating with the Picket was just something they couldn't tolerate and didn't want. And they would use that to try and come onto the picket and infiltrate, harass, or provoke people into reactions and stuff like that. And equally they made a lot of sexual remarks about, directed at women and again they're to provoke, they're to wind people up and get them to react and anything that they could use to offend people and wind people up and try and get the Picket disrupted really.[5]

In addition to high-profile political responses such as 1988's 'Hands Off Black Activists' campaign,[6] City Group responded collectively to protect black activists (and others) who were targeted by the police on the picket:

> Well very often it was just people who would really sort of gather round the people who were being assaulted or stand there with them and chant really loudly.[7]

For Ann Elliot, despite being involved in socialist and anti-racist politics since the 1960s:

> The Picket was a sharp learning curve and established ... procedures for dealing with the police, namely with good manners, knowing the laws are on our side and through Stewards, which made, bought the movement a step forward. The police were I think very surprised by the conduct of the Picket because, and in particular the protection that was given to black people, minority people, and I include here people who were homeless, or perhaps very needy, who lived on the streets,

and even to some extent gay people. And the Picket was protective and united in their defence of these people, and that was a big challenge to the police, they couldn't divide us in that way.[8]

As a black picketer, Tunde, who experienced repeated arrests and assault by the Police, recalled that City Group could protect individuals by encouraging them to contribute to anti-apartheid campaigning in another way for a while:

> The life in the office itself [was] nice, you get used to it after a while, this is actually better than standing on the picket line, you can drink as much coffee as you want... And this was actually important for me ... which is because too many arrests by the police, so I was told to maybe spend at least more than half of the time away from the Picket into the office. Getting arrested, because I became a target, you know, that was the reason.[9]

Young black men like Tunde became vulnerable to increased police racism by participating in the Non-Stop Picket and expressing their solidarity with those resisting racism and apartheid in South Africa. They were made 'vulnerable' by systemic structures of oppression, as much as their resistant stance. The solidarity that they found on the Non-Stop Picket helped them to deal, as part of a political collective, with police racism, but also with the other ways in which systemic racism and other forms of oppression affected their everyday lives. As Andre reflected, however vulnerable individual picketers might be in other aspects of their lives, the positive experiences and connections that time spent on the Non-Stop Picket offered to its participants could help them deal with those vulnerabilities:

> On the other hand we had this thing which is comradeship, which is a great experience, which we had and which was unique in Thatcher's time...it was a rare thing, and it's become even much rarer still, and yet it's great to have a sense of what that means and to be part of that. And even though what we experienced was nothing like a real struggle, you do get a taste for it ... You do have a better sense of, just on the human level, what it's like to go through something and have people there supporting you, waiting for you at the police station or going through a trial for political reasons. And that's unique, nowadays especially, there's not that many young people really that go through any form of political campaigning, and it's great to be part of it.[10]

From the South African struggle, Norma Kitson had introduced into to City Group the practice of referring to all participants as 'comrade' – a marker of respect and radical commonality. Comradeship was there in the ways City Group built collective solidarity and forged through direct action and confrontational practices with those resisting apartheid. It was there in the ways City Group responded collectively and politically to all threats to their ability to maintain a 'non-stop' protest for Mandela. It was there in the way that the pace, longevity, and intensity of their non-stop protest fostered unexpected friendships and practices of care between picketers across the boundaries of social and political difference.

For Cat, who joined the Revolutionary Communist Group (RCG) during her time on the Non-Stop Picket, it was important that this ethics of support and solidarity transcended differences of political perspective:

> And I mean obviously people who weren't in the RCG were also my comrades and that sense of solidarity with people who shared your cause I think was extra-ordinary…In fact generally it didn't matter whether you liked or didn't like people, you defended them against the police. You went to the police station if they were arrested, you supported them publicly, you could have your political arguments elsewhere, and I think that sense of collective solidarity was important.[11]

Politically, the Group knew how to respond to threats from the police and developed a variety of practices to either diffuse those threats or confront them. It was trickier to respond to other threats to (potentially) 'vulnerable' picketers. Sometimes these threats came from within the Group and its allies; at other times, they were incidental to the form and location of the Picket as a continuous protest on the streets of Central London, day and night. Ruby disclosed an incident when she was propositioned by an exiled member of the African National Congress (ANC) she had met on behalf of the Group. (Other picketers alluded to other, similar incidents, some of them very serious). Describing her experience, Ruby acknowledged how young picketers' political commitment to the anti-apartheid cause contributed to making them vulnerable in such circumstances (see Suttner 2008, Chapter 6 for a discussion of the problem of sexual abuse within the ANC's exiled hierarchies):

> Because you held them in great respect and a bit on a pedestal really, and I can't remember his name, but I remember he'd arranged to meet me somewhere like Finsbury Park. He basically propositioned me, you know, and there were lots of young girls. We were all kind of young and probably a bit naïve really because it was the ANC, freedom fighters and all of that. So it was pretty sleazy really, you know.[12]

In recalling such incidents, some former picketers reflected on how much policy discourses (and practices) around 'safeguarding' children and young people have changed since the 1980s. Maggie, a social worker, reflected on whether she would find it as professionally comfortable to stand alongside 'vulnerable' youth (without intervening 'professionally') within today's child protection and safeguarding frameworks.[13] She recalled, at length, an incident where she, Ann Elliot (a teacher), and some of the other, older, and more politically experienced women on the picket confronted a man they suspected of attempting to groom young picketers. She acknowledged that some of the younger picketers questioned this intervention, believing that it compromised their autonomy as young people:

> Ann went bananas, she was brilliant. And I was like, yeah. But some people, like younger people, were more like no, you know, like, they were misguidedly thinking this was a liberation issue or a children's liberation issue. Because that's how the Paedophile Information Exchange people put it that children had a right to sexual relations with adults and things like that ... Ann was absolutely stalwart on that one ... Ann was very good and we saw them off ...certainly we made a plan and ... all the young people were told not to, and he was spoken to. Somebody took him aside and ... I mean there was no violence but he was actually pointed out, and I think what we did was make it public and point him out, and have his photograph maybe, probably in Steward's book or something there was a thing saying he is not allowed on the picket and to warn young people. And for a while there was a lot of consciousness about it to make sure to always point him out and always make sure that particularly young people that might have been looking for a bed for the night or, that's how they pick them up. [Interviewer: What age were the young people involved?] 16, 17, 18. Maybe 19. Maybe 20 ... Not talking about children, definitely not. They weren't like 10-year olds, no. But it was young vulnerable people. So paedophile's the wrong term to use. But he was probably like a pimp.[14]

The Picket's location in Trafalgar Square connected it to the social and sexual geographies of the West End of London (Mort 1998; Hubbard 2013). The Non-Stop Picket created an agonistic political space in which children, young people, and adults could all participate. The 'vulnerability' of some of those young people (as well as the adults) did not usually impede their ability to be politically engaged. Indeed, many found that the comradeship that they experienced through their engagement provided them with support and solace, and sometimes facilitated them to take control of finding solutions to their vulnerabilities. The significance of this is not just that the web of comradely relations and practices of care provided support to the vulnerable. By being enrolled, as comrades, in providing care and solidarity to others, vulnerable young people were afforded the opportunity to transcend their own vulnerability and be strong for (and with) others.

Even for those who were not 'vulnerable' in some way to begin with, the pressures of maintaining a continuous protest over an extended period of time, in frequent conflict with the Police and other state agencies, could make some picketers vulnerable in new ways. Several interviewees reflected on their own and others' experiences of exhaustion and 'burnout' during the Non-Stop Picket (Cox 2010). Many activists were susceptible to these pressures, but they were exaggerated the more intensely people committed to the politics of the Picket and the task of keeping it going at all costs. Mark Farmaner observed:

> ...When people get sort of tired or a bit burnt out they also can become very destructive and my god did we have lots of that going on. Some people didn't just think I need a break or to leave; they wanted to bring everything down with them. I think you've seen with politicians what they call the Westminster bubble. There was a City Group bubble were people got a totally inflated idea of everything and their resignation was going to shake the world, you know, and it's like no one ever saw them again and they were forgotten about.[15]

Mike Burgess, who, having been a member of the City Group committee for several years, found himself crashing out of the Group amidst accusations of disloyalty and ill discipline, made a similar observation in relation to how and why some picketers started talking to the Police, against the agreed rules of the Picket:

> People who began as decent serious people, got involved and just got tired, just got to the point of fatigue where they couldn't carry on, and how do you walk away, how do you leave when you can't go on? And I felt that on several occasions you would see somebody go and cross a line, and talk to the cops and join the enemy knowing that you can't walk back to the picket now, you've gone, you've crossed a line, you can't cross back again. And I sometimes felt that that was somebody who just needed a way out, and even though they did the unspeakable and they gave away information to the other side, and they put us in a difficult situation, very often I believe that may have been because they simply got to that point where that was the only way they could quit.[16]

Clearly, the pressures of being a 'non-stop' anti-apartheid campaigner could be exhausting. Despite (and sometimes as a result of) the intensely close friendships and networks of comradely care that existed on the picket, the commitment to maintaining a non-stop protest until Mandela was released could significantly impact activists' mental and physical health. For some participants, especially those lacking resilience (Masten 2001) or the capacity for proper self-care, the only viable way out of these pressures was to alienate themselves from the comradeship that had held and sustained them within the Group. These issues continue to trouble activist campaigns to this day (and despite the many positive aspects of its culture of comradeship, is not something that City Group ever satisfactorily resolved). There are, however, many positive lessons that can be taken from the Non-Stop Picket, which we examine next.

Lives since the Non-Stop Picket

The experience and intensity of being 'non-stop against apartheid' left an indelible impact on the young people who stood in all weathers outside the South African Embassy in the late 1980s. The Picket generated lasting friendships and shaped the intimate and family lives of some participants. Former picketers identified tangible ways their involvement shaped the choices they made about their education and careers, subsequently. For some, the legacy of the Picket was not easily articulated. They knew it had touched them intensely and affected the way they behave to this day, but struggled to describe how it had done so. Many picketers could trace the ways in which time on the picket had cultivated political beliefs they continue to adhere to. But, a minority recognised their time on the Non-Stop Picket as a period of exception, a teenage adventure that (although cherished in various ways) had little direct impact on the course of their adult lives. Here, we examine ways in which picketing the South African Embassy shaped the lives of participants over the subsequent 25 years. First, we focus on the personal and professional legacies, then the lasting political impact on the lives of non-stop picketers.

For several picketers, their time on the Non-Stop Picket shaped their lives in real and tangible ways; they became parents with partners they met there. Charine explained:

> [I] met someone from the Picket in 1987. We got married, had a child in 1989 and about 1992 we moved back to Sierra Leone where we stayed for a few years and came back, separately, split up out there, and came back after a couple of years.[17]

Andre was from a younger generation of picketers, but had a similar story to tell:

> I've got obviously my son; his mother is also a former City Group activist so I wouldn't have had my son to start with. It's shaped me from a teenager, I was 16 when I joined the Picket, so that's from quite a young age.[18]

Picketers formed serious and long-term relationships, several of which have lasted to this day – we interviewed at least two couples who first met outside South Africa House. Interviewees said they had remained exceptionally close friends with former picketers.

Deirdre was not alone in assessing that the depth of the friendships she has with some other City Group veterans, and the density of the connections between those friends, was "near family or near, I mean it *was* family."[19] She offered several examples of the ways in which friends from the Picket (now dispersed across several countries) had gone out of their way to help her years later.

It is a measure of how intense the friendships were between picketers that some of the people we interviewed acknowledged that subsequent friendship networks were never able to live up to their standard. For example, Gary Lowe reflected on a difficult period in his life soon after the end of his involvement with the Non-Stop Picket:

> I mean thank god there was uni to give me kind of stability, but the people at uni were not like the people at the picket.[20]

Similarly, as Deirdre accepted, few subsequent political campaigns and protests could offer the same intense and exciting atmosphere that impelled her to be involved:

> I haven't found anything that quite lights my spark as much as that did, and I'm ashamed to say that that might be age related now that I am nearly mid-forties.[21]

Deirdre's reflection implies that there was something in the qualities of the Picket that combined with her age at the time to encourage her commitment. Articulating what that 'something' was could sometimes be tricky for former picketers, just as they could not always clearly articulate how the experience of being a non-stop picketer had left traces in their attitudes, dispositions, and personal qualities.

Jacky Sutton suggested that the personal qualities she had developed through her involvement in the Non-Stop Picket had later been tested to the limit through her experiences as an aid worker in conflict zones.[22] Notably, those who had joined the Picket as young women spoke most directly about how that experience had taught them to be 'stronger' and more determined (and able) to defend their opinions (Taft 2011). These were qualities developed by many picketers, but they especially challenged normative understandings about how young women should act in public (cf. Roseneil 2000). Here's how Irene put it:

> It made me unafraid of conflict. I think probably on a personal level that's probably the biggest thing it gave me. It gave me a sense of loyalty and commitment to people and to campaigns and seeing something through and so on, but I think in general terms I was a schoolgirl and I was quite a shy schoolgirl and I think to put yourself in that situation you learn a lot quite quickly and it toughens you up quite a bit![23]

Standing on a pavement in Trafalgar Square defending contentious people, ideas, and tactics 'toughens you up'. Like Irene, Cat also recognised that this 'toughening up' and losing fear of conflict did not occur individually and in isolation; it was a quality that young women (and their male peers) developed together through "loyalty and commitment to people." As Cat put it, she learned:

It's worth being disciplined, it's worth a sense of comradeship and collectivity, I mean there were those sorts of lessons I think that were important, you know, apart from the standing on the snow or the standing in the rain or standing in the middle of crowds singing Auld Lang Syne and trying to sing Nkosi Sikelele instead on New Year's Eve or in that storm of 1987.[24]

As someone who has dedicated herself to the collective discipline of a communist organisation over the intervening decades, it is perhaps not surprising that Cat stresses the importance of these lessons. Penelope celebrated the ways that experience had "reinforced my rebelliousness and made me a little more disciplined and assertive."[25]

It was not just young women that experienced a new determination and a new mode of engaging with the world as a result of participation. Nick Manley was at least a decade older than many of the women quoted above when he broke from his religious fundamentalist upbringing and joined the anti-apartheid cause:

It gave me, eventually (after all the sifting), a coherent take on the world and everything that involves, really – the politics; I've got a lot more tolerant of different lifestyles. Rightly or wrongly, it has given me a greater understanding of the world I live in, and a more realistic take on what's worth doing to change anything.[26]

Many of the people who had joined the Non-Stop Picket as children and young people outlined ways in which the political focus of the campaign, as well as the practical skills and dispositions acquired through it, shaped their careers as adults. For Deirdre, her political interest in apartheid shaped her decision to work in international development education and training – work that has taken her to Southern Africa a number of times.[27] Ben, who worked as a theatre director and for the British Council, acknowledged that "I do lots in South Africa now, with that cultural interest."[28]

Throughout this book, our analysis has looked beyond easy tropes of adolescence as a time of experimentation and identity flux to consider how growing up under the Thatcher government profoundly changed the ways young people thought about their futures and their social and personal identities. As Selman reflected:

The Picket wasn't the start, the start happened before then, but it was certainly a galvanisation of a path ... I think it is a very, very important chapter in my life. Half because of the time of my life that it happened, it happened at a very formative stage, and half because of what happened during that time. You know, there was a lot of discussion, a lot of activism and a lot of social events that stayed with me, if not in their detail, [then] in the fact that they happened.[29]

While many picketers remain politically active (as we discuss later), a few have chosen to 'professionalize' their campaigning skills. Mark Farmaner is one of them:

> I'm still campaigning now 20 years later ... What I did was campaign and I wanted to keep campaigning and so I did that unpaid and voluntarily for a long time before I accidentally got a paid job campaigning at Christian Aid.[30]

Mark acknowledged that having spent his teenage years committed to the Non-Stop Picket at the expense of formal education and training, he had missed out on developing other skills that might have opened up different career opportunities. The Non-Stop Picket provided picketers with a chance to develop certain skills and dispositions, but these could also create particular 'path dependencies' in their lives.

For many former picketers, the influences of formative years spent campaigning outside the South African Embassy were all encompassing. The Picket's influence could not be contained within a single aspect of their lives. Helen Marsden got very emotional as she elaborated on the numerous ways the Picket lived on in her life. She was determined to demonstrate that her beliefs and commitments were not 'just a phase' from her teenage years:

> I'm still in that phase ... But it has shaped my whole outlook. It's the reason I don't shop at Tesco; it's the reason I've got a picture of Camilo Cienfuegos[31] up [on the wall]; it's the reason why Public Enemy is my favourite band; it is the reason I spent 11 years in Cuba. It's the reason that I explain the world to my children. I feel quite emotional. I do, oh goodness. It was really a big part of our lives.[32]

Although Helen spoke at length about the influence on her own life, she understands the Non-Stop Picket as a collective experience – "it was really a big part of *our* lives."[33] Those beliefs, commitments, and choices were shared with others. Like Helen, Charine expressed her on-going commitment to "street-level political action and non-violent direct action" as a shared commitment held in common with others, "that we have far more things in common than we have that keeps us apart."[34]

Nevertheless, from the perspective of early middle-age, several picketers acknowledged that the Non-Stop Picket happened at a point in their lives when they had an excess of free time, which enabled them to commit to it. As Nicole said:

> I've always been politically interested. I've never had as much time on my hands as I did then as a student. I feel I was really quite lucky at that point in my life not to have the responsibilities that come with adulthood – jobs, houses, etc.[35]

Sharon makes a similar reflection:

> I didn't stay involved after the Picket ended. I started training to be a nurse.... I've been living outside London and bringing up three children, which, while not an excuse, has made it difficult for me to take part in any political campaigns/demonstrations. My views are unchanged and I will I am sure, be more active again in the future.[36]

Grace described how the Non-Stop Picket had "really shaped" her life in the sense that through it she "became completely politicised and I've remained political ever since."[37] A remarkably high proportion of our interviewees continue to lead highly politicised lives, shaped by on-going (if not always continuous) political activism over the quarter of a century since Mandela was released and the Picket ended. Below, we focus on their continuing activism and the ways it has been shaped by their experience of being 'non-stop against apartheid'.

Politics since the Picket

Some non-stop picketers brought decades of political and campaigning experience with them; but, for many, it was their first real taste of activism. This book focusses on the experiences of young picketers; and, while they were more likely to be cutting their political teeth outside South Africa House, not all of them were entirely politically naive. Even some of the teenage activists had already accumulated campaigning experience (in the anti-nuclear movement, in animal rights campaigns, miners' support groups during the 1984–1985 strike, or in pre-Non-Stop Picket campaigning against apartheid, for example). Whatever level of campaigning experience they brought to the Non-Stop Picket, those four years of continuous anti-apartheid protest and solidarity work profoundly impacted their subsequent political campaigning. Just as the picketers came from a diverse range of political backgrounds, so their campaigning followed various trajectories after the Picket ended. Nevertheless, there are a number of patterns that emerge: some picketers transferred their experience of international solidarity campaigning to other causes in other countries; some applied the practical campaigning and political organising skills honed on the picket to local and national causes; and finally, through their informal political education during long hours on the Non-Stop Picket, others became convinced anti-capitalists and committed to revolutionary politics of different complexions.

Alongside their anti-apartheid campaigning, many City Group supporters (influenced by the arguments of the RCG, although not always entirely in agreement with them) were also involved in Irish solidarity work, demanding British troops leave Ireland, in sympathy with the Irish Republican movement (McKinley 1991; Reed 1984). Lorna Reid was among a number of former picketers who chose to put their energies into this solidarity work immediately after the Non-Stop Picket ended. Despite her long-running commitment to anti-apartheid campaigning, she recalled:

> I remained in Irish politics. That was always my first love, if you like. So I was active in that.[38]

Through the influence of the RCG, a number of erstwhile City Group members also became involved in international solidarity work with socialist states in Latin America – first, offering practical and political support to Cuba, against the US Blockade and in support of the socialist revolution; and later also in solidarity with Venezuela (Menendez 2013). The RCG set up the solidarity campaign Rock Around the Blockade (RATB) in 1995, the year after City Group finally ended. It was founded on many of the political principles and activist approaches that were central to City Group's campaigning. It offered more than 'humanitarian' aid, emphasising solidarity that flows in more than one direction; using the example of Cuba to build a socialist movement in Britain. Kathy was one of those who explained that, after the Picket, she "went on a couple of brigades to Cuba."[39] These annual solidarity brigades to Cuba are opportunities for brigadistas to learn about the challenges and achievements of Cuban socialism through visits to different institutions, centres, facilities, and to learn how society is organised through meetings with political, workplace, and neighbourhood representatives. The brigades provide material aid for Cuban youth (through the Union of Young Communists (UJC) and, for the first 15 years, donated expensive sound systems and disco equipment for Cuban youth, all of which was fundraised for politically). Although RATB has never organised a 'non-stop picket', it has engaged in similar activist practices that formed City Group's contentious repertoire: visually appealing lively protests and direct actions; educational meetings, relating international solidarity with the struggle for rights in Britain. This is not a coincidence; several key activists from the Non-Stop Picket – Cat, Hannah, Ken, and Richard – set up and led RATB for many years, and others were also involved.[40] Additionally, members of the RCG organised weekly Palestine solidarity protests around the country. The one on London's Oxford Street was maintained for 14 years – again reproducing many of the practices and organisational methods developed through City Group's anti-apartheid campaigning.

Andy Higginbottom, who had been City Group's Secretary at the start of the Non-Stop Picket, led an international solidarity campaign first in Kurdish solidarity work and then as the Secretary of the Colombia Solidarity Campaign supporting trade unionists and human rights defenders in that country.[41]

For several years, Mark Farmaner, who played a significant role in City Group, has been the Director of the Burma Campaign UK, working in solidarity with human rights advocates and in defence of ethnic minority groups in Myanmar. While the experience of organising the Non-Stop Picket influenced Mark's subsequent campaigning, he also recognised the gaps in his campaigning skills that he had to fill in later roles. He explained that City Group's particularly confrontational style of campaigning did not work in the context of his current work for the Burma Campaign UK.[42] Although City Group was reasonably good at engaging sympathetic journalists and politicians, Mark argued that in his subsequent campaigning work, he had needed to develop new analytical and public relations skills in order to influence a wider range of policy and opinion makers. He had had to learn to be more 'diplomatic' (Dittmer and McConnell 2015).

One international solidarity campaign from the 1990s that, despite considerable differences from the anti-apartheid cause, was partly inspired by the experience of the Non-Stop Picket was Workers Aid for Bosnia. As the civil war in the former Yugoslavia deepened in 1993, a small group of British socialists, trade unionists, and Bosnian refugees met in London to consider how they could offer support to working people in Bosnia in the face of the Serbian invasion and ethnic cleansing (O Tuathail 1999; Jeffrey 2012).

One of the organisations central to the formation of Workers Aid was a small Trotskyist group, the WRP (Workers Press), who had split from the larger Workers Revolutionary Party in 1985 following allegations of the sexual abuse of women members by the party's leader Gerry Healy (Burton-Cartledge 2014). Several members of the WRP had been involved with City Group since the 86-day picket for David Kitson in 1982, where they provided night security (Kitson 1987). A few WRP members stayed involved with City Group over the years, and most of them were amongst the people who left to form the Workers Press group in disgust at the political culture within the larger party that had allowed Healy's abusive behaviour to go unchallenged. One of the Workers Press supporters who attended the founding meeting of Workers Aid for Bosnia was Bob Myers who had been City Group's Trade Union Officer. He described how he saw the similarities between City Group and Workers Aid:

> A few years later I got involved in organising convoys of aid to Bosnia during the war in Bosnia, and of course we ended up with very similar mix of people. Ex-soldiers, unemployed people, political activists, ex-political activists, a very similar kind of mixture of people motivated by the most thought-out political ideologies and people who just thought it would be a laugh, you know. So that in a way was good training because later on I got involved in very much organising the convoys, and having to deal with all these problems that you get when you have a kind of thing like this involving such a range of different people with different motives and different interests.[43]

One of the most direct ways in which Bob (and others from Workers Press) brought their experience of City Group into the solidarity work of Workers

Aid for Bosnia was when the Bosnia Solidarity Campaign organised a continuous protest opposite 10 Downing Street and the Foreign Office, which lasted from 19 July to 29 September 1995. They even called it the 'non-stop picket' (Workers Aid for Bosnia, n.d.: 133).

Other picketers used skills they had developed with City Group in a range of other campaign settings in the 1990s and 2000s. Lorna Reid went on to stand, unsuccessfully, in the 2004 London Mayoral Election on a platform of independent working-class politics[44] (Hayes 2014). Many, like Lorna, had been active in the movement against the Poll Tax (Bellamy 1994; Burton-Cartledge 2014) prior to the end of the Non-Stop Picket and sought to be active where they lived. Indeed, the South African picketer Shereen remembered the Poll Tax campaign as

> one of the first times we became involved in an issue which was British rather than the other way around. And yeah, it was interesting for us to see how British people campaigned for themselves. We hadn't been here for the miners' strike or Wapping; we'd come just after that, so it was, yeah, it was interesting.[45]

Several former picketers expressed frustration that they had neither found a cause to 'replace' the Picket in their lives, or that their experience was not always welcomed by the new activists they attempted to work with. Andy Privett spoke enthusiastically about the connections he had made in recent years through his involvement in UK Uncut and Unite Against Fascism. However, he questioned the apparent reticence of some contemporary activists to replicate even the more minor acts of direct action (such as 'trolley pushes' in supermarkets; see Chapters 6 and 7) that were staples of City Group's agitational practice in the 1980s.[46] Similarly, Penelope Reynolds complained about fellow Palestine solidarity activists in Bristol:

> They won't listen to my advice about how to mobilise hundreds of young activists, and that's the lifeblood of any kickass campaign ... I suppose there aren't any embassies here in Bristol and that does make it less effective to act local in solidarity with overseas activists.[47]

After the intense collectivity of the Non-Stop Picket, many of those who continued to do political work, as individuals, complained that they had become "politically frustrated."[48]

In contrast to these individuals, other former picketers have continued their activism as part of organised political groupings. There were organised groups of communists (mostly from the RCG, but also Workers Press) involved in City Group. Even though some of their members had years of political experience prior to the Non-Stop Picket, it still had a profound influence in shaping the development of their political ideas and practices. Carol Brickley asserted:

> It was the most important formative political experience of my life and I see it as a political experience, not as a personal one. I learnt a lot. And I had the opportunity to mix with some amazing people who had given their all, their lives, to the struggle. That's a privilege to see that. Those things that happen are important for the consciousness of any movement; they become part of its history and its future. They aren't lost. Those victories aren't lost. The City Group experience is a starting point, rather than the end point. The next movement will incorporate that knowledge and that experience. It's important to pass it on and I'm glad that it's being passed on.[49]

The political leadership that Carol and other members of the RCG provided within City Group convinced some young picketers of the veracity of their communist critique of contemporary capitalism. The RCG never dominated City Group, numerically or politically, but its ideas were influential. Over the years, a number of young picketers joined the RCG for a period of time, but only a minority stayed in the Group long-term. Nevertheless, there are a number of people who first came into the contact with the RCG through City Group and remain active with the RCG to this day. For them, this contact has profoundly shaped their world view and political activity since then. Cat and Hannah told very similar stories in this respect:

> Well, it brought me into contact with communists, I joined a communist organisation, and that's probably over the last 26 years influenced how I lead my life. I mean I lead it as part of a communist organisation and I see the world in that way.[50]

> Well if it wasn't for the Picket I probably wouldn't have met the RCG and that's definitely changed my life, because it occupies a lot of my time ... My involvement and understanding of world politics and ... my being part of an international struggle for socialism and defence of Cuba and Revolutionary Latin America, and [the] Bolivarian Revolution in Venezuela, all of that is as a result of my involvement with City Group.[51]

Debating politics with members of the RCG helped convince a wider layer of picketers of an anti-capitalist critique, even if ultimately they chose to express this through involvement with other forms of radical and revolutionary politics. We spoke to former picketers who had joined the Class War Federation in the early 1990s (Franks 2006) and others who had joined the Revolutionary Communist Party (which originated as a split from the RCG in 1976 over the political significance and role of Third World national liberation struggles). Although the RCG played a major role in convincing many picketers of an anti-capitalist politics, they were not the only figures associated with City Group who had this kind of influence. Mike Burgess explained that, after he left City Group, the organisation he decided to join was inspired by his respect for the politics of David Kitson:

> I mean I didn't understand his ideology but I respected it. I respected it so much that I began reading *The Leninist* because I heard that he read *The Leninist*, whatever, there were letters in *The Leninist*. And so that drew me to joining that organisation, simply out of respect to David Kitson.[52]

Clearly, former members of the Non-Stop Picket have gone on to contribute to and shape many high-profile political campaigns on the British Left over the last 25 years. In describing her own political trajectory, Lorna Reid noted:

> I was politically active before the Picket, and then I was politically active after it. I think you pick up lessons and skills from whatever you do, and there was lots of lessons and skills to be gained from it.[53]

We concur. While it is historically interesting to trace the subsequent political trajectories of City Group activists following the Non-Stop Picket, it is also instructive to examine the skills they carried forwards with them and could apply and rework in new contexts. This evaluation will highlight how the form of solidarity performed on the Non-Stop Picket had an enduring effect after the protest ended.

Reusing competencies; reworking practices

It is possible to view the Non-Stop Picket in various ways: as a protest event (albeit one with an extended, elongated duration); as a protest site; or as a set of protest practices. Each of these ways of thinking about the Non-Stop Picket (temporally, spatially, or as a form of doing) is valid, but in isolation, none of them capture enough of what it was like to be there and to participate in being 'non-stop against apartheid'. We have focussed on thinking about the space created by the Non-Stop Picket (Chapter 3), the ways in which picketers sought to defend their territory (Chapter 5), as well as the ways in which time passed and was marked on the picket (Chapter 4). To understand the ways in which having been a 'non-stop picketer' has continued to shape aspects of their lives, it is most useful to consider how the Non-Stop Picket was assembled from a bundle of related practices (see Chapters 2 and 3).

In the light of former picketers' reflections on the ways in which the Picket has shaped their later (adult) lives, we can evaluate the component practices of the assemblage that was the Non-Stop Picket. Throughout this book, we have argued that being on the Non-Stop Picket involved practising solidarity (with people resisting apartheid in Southern Africa, with other anti-racists, and in the face of heavy-handed policing with each other). The core practice of (anti-apartheid) solidarity was in fact a bundle of other associated practices – organising, mobilising, fundraising, witnessing, debating, protesting, taking direct action, and others.

Extending the approach to social practices advocated by Shove et al. (2012), each of these practices can be understood as emerging out of the dynamic interaction of their different components elements: the materials (objects, things, and technologies needed to undertake a task), the competences (in the form of skills and know-how needed to do the task), and the meanings (attached to the significance to that action). The individual elements of any given practice are often elements in other, related, practices too. So, for example, as we explored in Chapter 7, for many young picketers (and even some older participants whose lives were in a significant transition for other reasons), the practice of performing solidarity on the Non-Stop Picket shared many elements with the practice of growing-up with others who shared their outlook on the world. Sometimes one practice can serve as an element of others. Key components of being non-stop against apartheid were rebelliousness (against authority), discipline (to a collective), assertiveness, and the ability to defend unpopular opinions. Key resources were also the ability to work collectively with people from diverse backgrounds and offer comradeship to each other. Comradeship, in this sense, might be understood as an element of practicing anti-apartheid solidarity, but it was also a practice in its own right.

When asked how the Non-Stop Picket had influenced her life, Grace said,

> I think it was a really important experience of seeing how people could get together, achieve something, and you know, achieve a lot. Basically, it was a really positive experience because it showed you how people can come together and really make a difference and organise a campaign that was really effective, that was fun, that was democratic. So it was a sort of model for you to try and follow in later life, and so I think, in that way lots of people have stayed political because it was a really inspiring campaign.[54]

Her answer is instructive, because she acknowledges that the experience of being on the Non-Stop Picket led her to acquire practical skills (Hitchings 2012), to 'organise a campaign that was really effective'; but the fact that this had occurred on the Non-Stop Picket gave these competences meaning and ensured that "lots of people have stayed political because it was a really inspiring campaign." Skills acquired in one context become transferred to a new context and take on new meanings. The competences might be the same, but they are transformed by the new (contextual) meanings that are attached to them.

214 *Lessons and reflections*

For Liz, involvement with City Group gave her

> a good basic education in how to run a picket. I can now certainly look at a picket and say nah! They're not doing that properly; they need a bit of training.[55]

This would appear true for many former picketers; but it can play out in a number of different ways. The Non-Stop Picket certainly provided picketers with a range of skills and know-how that they could transfer to later campaigns (and many have done this). However, at times, the failure of those other campaigns to practise solidarity, disobedience, and protest in exactly the way that were undertaken on the Non-Stop Picket can be alienating and upsetting. The symbolic association of those practices with the Non-Stop Picket has become so strong that their 'meaning' acts as a break on those former picketers being able to effectively rework the practices in new contexts. This explains the 'political frustration' acknowledged by some of the people we interviewed.

Mark Farmaner (like many other picketers) asserted that a key skill he carried with him from Non-Stop Picket days was a certain way of dealing with the Police (whether in the context of political demonstrations or just encounters on the streets). He said:

> I'm completely not intimidated by police and I stand up to them and challenge them all the time. I mean I would go out of my way to challenge them because of how they've behaved … It's immediate distrust and suspicion of police and they still, most of the time, they fulfil my expectations. They still behave in an illegal and arrogant way.[56]

However we choose to name this practice, it is a competence that Mark and others acquired on the Non-Stop Picket and has been given meaning for them by its historic association with that protest. This practice finds a very specific expression (and carries with it certain meanings) in the ways that former picketers still challenge the Police when they think they are acting inappropriately, over-extending their powers, or being unjust. However, it is at its root the same practice of adopting a stance of indignation mentioned in a variety of contexts by others – whether that is defending unpopular ideas, or challenging an undemocratic decision in the workplace. For Cat, this practice was fundamentally about learning to overcome embarrassment in front of others. She asserted:

> One of the things that was really useful I think for anyone who was involved in the Picket particularly in a country like Britain where people get a bit embarrassed about stuff and doing stuff in public, everyone who was there lost their fear of standing in public, handing out leaflets, making speeches, standing up to the police … [it was] that thing about

losing your fear of standing up for something, because of what people will think or because you'll look stupid or because you'll be the only one.[57]

While this stance came easily for some picketers, many others learned it through the Non-Stop Picket. Norma Kitson held workshops for young picketers on public speaking and how to use a megaphone effectively on a protest. Central to these classes was practicing overcoming embarrassment and fear. These are skills that many picketers continue to apply in all kinds of contexts in their lives today.

Picketers have carried with them an instinctive response to dealing with arrests in the context of political protests (see Chapter 6). None of these practices were invented on the Non-Stop Picket (Tilly 1999), but they became a central part of the practice of being 'non-stop against apartheid' and have carried particular meaning for many as a result ever since. As David Yaffe reflected (both in relation to his own campaigning and that of the RCG as an organisation):

> All these questions which I would see now as part and parcel of the movement I belong to – the RCG – it became clearer and clearer and developed significantly because of the campaigns in City of London Anti-Apartheid Group, no doubt whatsoever. So in that respect, my understanding, my political consciousness, the organisation's understanding and political consciousness developed enormously as a result of that experience.[58]

In the 25 years since the Non-Stop Picket ended, the global geopolitical context has changed significantly (Dodds 2005; Cohen 2015); so too has the domestic political situation in Britain, the legal framework for policing protests, and approaches to safeguarding young people (Kraftl et al. 2012). All of these changes and others alter the possibilities for reworking the particular practices associated with the Non-Stop Picket. They also alter the meanings attached to those practices. On the one hand, being against apartheid and supporting Nelson Mandela is celebrated today in a way that it never was in the 1980s. At the same time, the political concerns associated with young people's engagement with 'radical' politics are far more cautious today.

Former participants in the Non-Stop Picket have made the skills that they acquired through their engagement in radical politics into a resource that they have been able to utilize at later points in their lives. For example, Adam Bowles, who served time in prison for his part in throwing red paint over the front of the South African Embassy in protest at the white-only election in South Africa in May 1987 (see Chapter 5), even turned this experience to his advantage. His conduct during the trial so impressed his barrister that he was offered a job in chambers as a barrister's clerk.[59] In multiple, similar, small, and unexpected ways, the activism of youthful anti-apartheid protesters on the Non-Stop Picket went on to shape their future lives.

Former picketers continue to rework skills and practices they developed through their anti-apartheid campaigning in their personal, professional, and political lives. A great many of them continue to hold radical political views and engage in a wide variety of progressive political causes – from those seeking justice in London's housing market (Watts 2016) to those addressing issues of international and global concern. Not only are young activists' political commitments always entangled with the everyday politics of growing up, they are a positive resource for socially engaged adulthood.

Lessons of the Non-Stop Picket

Perhaps the primary lesson to be learnt from the Non-Stop Picket's longevity and success was: *choose your location carefully*. The Non-Stop Picket gained credibility and traction because it was located directly outside the South African Embassy. This was the point in London where maximum embarrassment could be caused to the representatives of the apartheid regime in Britain. In that sense, the siting of the Non-Stop Picket was strategic. An additional benefit was South Africa House's location in Trafalgar Square, in the centre of London. As well as a major tourist attraction, this area is a key node in the city's transport infrastructure. Consequently, the Non-Stop Picket was seen by thousands of people every day. This not only gave the protest visibility and publicity, it also gave it a source of recruits to sustain it.

Two further lessons follow from this. First: *use the location to your advantage*. City Group used the Non-Stop Picket's location on Morley's Hill on the east side of Trafalgar Square to their advantage, symbolically and materially. Several of the Group's best orators (Trevor Rayne, in particular) were keen to locate the Non-Stop Picket within the long tradition of protest movements who had used Trafalgar Square as a focal point for demonstrations in support of democracy, universal suffrage, and a range of progressive causes since the 19th century. This symbolic association was particularly important in 1987 when City Group had to fight to defend their right to protest against apartheid directly outside the South African Embassy. More materially, the Group frequently used the width of the pavement outside the Embassy to their advantage in court, when defending protesters against charges of highway obstruction or similar offences. Equally, the background noise of the traffic circulating Trafalgar Square was key to defeating charges of 'noise pollution' that had been brought to prevent the Picket from using megaphones to amplify their demands during the Embassy's opening hours.

While the Non-Stop Picket showed no respect for the 'peace and dignity' of the South African Embassy (demanding it be closed down until the end of apartheid), they embodied a further lesson relating to their location: *be sensitive to those whose space you might be occupying* (Brown et al. 2017). They strove to maintain good relations with the parish of St Martin-in-the-fields

Church. More significantly, the Non-Stop Picket respected the fact that they shared Trafalgar Square, around the clock, with a local street homeless population. Although the Picket's interaction with homeless people in the area was not without tension, it did attempt to develop a symbiotic relationship with them. For those who could (be seen to) comply with the Picket's rules against alcohol and drugs, it provided a space where they could find company and be integrated. There were many times, when the Picket's rota was under strain, that the additional ad hoc attendance of homeless people helped to keep lonely picketers company too.

This exemplifies a broader, enduring lesson of how the Non-Stop Picket continued for four years: *make protests vibrant, lively, and encourage active participation from anyone*. The infrastructure of the Non-Stop Picket was usually quite minimal and basic, but it was used effectively (most of the time) to convey its message and make the protest look attractive. There was always a colourful, visually interesting banner held aloft. These were double sided so they could be seen and read both my passing pedestrians (on the inside) and passing motorists and bus passengers (as well as tourists visiting Nelson's Column) (on the outside). Frequently, picketers would wear hand-drawn placards and posters around their necks, adding to the visual display while keeping their hands free. The Non-Stop Picket also sounded exciting. Throughout the day, picketers would regularly use a megaphone to make speeches about apartheid and the anti-apartheid cause. Even a small crowd could sustain lively chanting of rhythmic anti-apartheid slogans or sing South African freedom songs (often accompanied by dance moves). Visually and sonically, these practices could capture the attention of passing tourists and members of the public. Picketers with petitions would then seek to engage these passers-by in conversation, to elicit their support and encourage them to stay a while or get involved. Because picketing the Embassy was an active, lively practice, there were always simple activities that newcomers could be encouraged to join in with so that they felt useful and part of the Group.

The Non-Stop Picket welcomed the participation of anyone opposed to apartheid and racism. The threshold for participation was low. City Group attempted to make the Non-Stop Picket not just welcoming, but an inclusive space, and created structures to encourage and support the participation of women; lesbians and gay people; black people; as well as youth and students. City Group recognised that some of the people who were most able to commit time to the Non-Stop Picket (and the Group's broader campaigning) were unemployed or on very low incomes. It therefore *facilitated the engagement by poor and low-income people* (paying their transport costs from public donations).

In many ways, the way in which the Non-Stop Picket operated stood in contrast to the functional anarchism found in some of its contemporary protest camps (such as Greenham Common) and many others that have occurred since. A key lesson of the Non-Stop Picket, then, might be to value *the importance of being highly organised* and not to reject hierarchical forms of organising out of hand. At various scales, from the individual shifts on the Picket's rota to the practical and political leadership of its campaigns, City Group organised hierarchically. Its leadership was accountable, elected, and (most of the Group's key posts) rotated over time, rather than being entrenched. Leadership was provided by the most engaged and committed activists at any given time. City Group was successful in giving responsibility for decision-making and organisation to young people (alongside those with longer experience). This inclusive leadership helped to foster collective notions of discipline within the organisation, and to achieve buy-in for a democratically agreed set of rules by which the Non-Stop Picket operated.

Key to the Non-Stop Picket's success was its practice of appointing a Chief Steward on every picket shift and every protest who could provide political leadership and to control interaction with the Police. The Non-Stop Picket actively educated participants about their rights to protest and how to conduct themselves if they were arrested. City Group fostered an understanding of the Police Officers as representatives of the state, rather than individuals. This helped ensure that the Non-Stop Picket, and individual picketers, kept a focus on their tasks as political and oppositional. This disciplined, political approach aided the execution of their protests and the legal defence of those who were arrested or targeted during them.

The hierarchical organisational structures in City Group were accompanied by *maximum internal democracy* in the weekly membership meetings. Here, the focus was on informal political education and discussion as much as practical organisational tasks. City Group welcomed the active participation of all individuals and political tendencies who were opposed to apartheid. There was *no censorship of minority tendencies*, and all participating groups were allowed to distribute their literature at City Group's events. These are important practices that are rejected by many contemporary campaigns to their own detriment.

This open and inclusive mode of organising was mirrored in City Group's approach to anti-apartheid solidarity. City Group was adamant that it should offer *non-sectarian solidarity* to all anti-apartheid tendencies in South Africa and Namibia. While, in practice, the Anti-Apartheid Movement (AAM) supported the ANC's promotion of itself as 'the sole legitimate representative' of the South African people (Thomas 1996), City Group recognised that AZAPO, the PAC, and others were also playing a key role in the struggle against apartheid and warranted the support of international solidarity campaigns too. This orientation to international solidarity took seriously that the anti-apartheid struggle was a battle for national self-determination by the majority of the South African people, and only they should choose their own leaders. City Group understood its *international solidarity as a political struggle* (rather than in charitable or humanitarian terms). The Group recognised that Britain had long-standing political, economic, and cultural links with South Africa that operated along 'long chains of command' (Massey 2008: 323) and that these needed to be exposed, contested, challenged, and reimagined politically. In this way, City Group sought to *make international solidarity relevant to domestic struggles*, linking the struggle against apartheid in South Africa to anti-racist and working-class struggles in Britain. City Group sought to mobilise supportive South African exiles to support anti-racist struggles in Britain and to give evidence, about the brutality of apartheid, in support of arrested campaigners in court. In this way, the Group understood solidarity as a political relationship that flowed in both directions between Britain and South Africa (as well as other nodes in international anti-apartheid networks).

Inevitably, this political and contentious approach to anti-apartheid solidarity resulted in many campaigners being arrested over the years. In response, City Group quickly learned to *defend protestors through political campaigns and active legal support*. They made legal defence procedures political and engaged progressive legal experts to support their approach. They used arrests as evidence of 'British collaboration with apartheid' to boost their political message. The robust nature of their highly organised legal support meant that more of their supporters were willing to risk (repeated) arrest than might otherwise have been the case. When their right to protest was constrained or put under threat, City Group was prepared to use a combination of political campaigning, civil disobedience, and the courts to defend their protest and its participants.

A final, but vital, lesson that City Group and the Non-Stop Picket can offer contemporary campaigners is: *do what works; don't limit yourself to what is respectable or acceptable to establishment norms.*

Notes

1 'We are here until Mandela is Free. Demonstrate 14 March 1987' (1987), *Non-Stop News*, No. 16, 23 January, p. 1.
2 'March for Mandela', *Non-Stop News*, Anniversary issue, April 1987, p. 6.
3 Messages of support were also received from overseas: from the Dunnes Stores Strikers in Dublin (Guelke 2000; Lodge 2006), sacked for refusing to handle South African goods three years earlier; and the Free South Africa Movement in California.
4 'Necklacing' was the use of burning car tyres hung round the necks of those believed to have collaborated with the apartheid regime during the township uprisings of the mid-1990s (Ball 1994).
5 Interview with Ken Bodden.
6 A City Group leaflet, *Hands off Black Activists: defend the non stop picket from police harassment*, August 1988 claimed: "from the middle of June to the middle of August the Metropolitan police launched a renewed offensive against the Non-stop picket ... black people have been the main target of the police's latest vendetta ... Seven black people arrested in eight weeks on a total of 15 charges! The Metropolitan police are out to stop black people in Britain for fighting in solidarity with their sisters and brothers in South Africa.'
7 Interview with Ken Bodden.
8 Interview with Ann Elliot.
9 Interview with Tunde Forrest.

10 Interview with Andre Schott.
11 Interview with Cat Wiener.
12 Interview with Ruby Noorani.
13 Interview with Maggie Mellon.
14 Ibid.
15 Interview with Mark Farmaner.
16 Interview with Mike Burgess.
17 Interview with Charine John.
18 Interview with Andre Schott.
19 Interview with Deirdre Healy.
20 Interview with Gary Lowe.
21 Interview with Deirdre Healy.
22 Interview with Jacky Sutton.
23 Interview with Irene Minczer.
24 Interview with Cat Wiener.
25 Interview with Penelope Reynolds.
26 Interview with Nick Manley.
27 Interview with Deirdre Healy.
28 Interview with Ben Evans.
29 Interview with Selman Ansari.
30 Interview with Mark Farmaner.
31 A Cuban revolutionary who fought alongside Fidel and Raul Castro and Che Guevara in the Cuban Revolution of 1959.
32 Interview with Helen Marsden.
33 Ibid (emphasis added).
34 Interview with Charine John.
35 Interview with Nicole Aebi.
36 Interview with Sharon Chisholm.
37 Interview with Grace Livingstone.
38 Interview with Lorna Reid.
39 Interview with Kathy Fernand.
40 Interview with Cat Wiener; interview with Hannah Caller; interview with Ken Bodden; interview with Richard Roques.
41 Interview with Andy Higginbottom.
42 Interview with Mark Farmaner.
43 Interview with Bob Myers.
44 Interview with Lorna Reid.
45 Interview with Shereen Pandit.
46 Interview with Andrew Privett.
47 Interview with Penelope Reynolds.
48 Interview with Daniel Jewesbury.
49 Interview with Carol Brickley.
50 Interview with Cat Wiener.
51 Interview with Hannah Caller.
52 Interview with Mike Burgess. For the origins of the group who produced *The Leninist* newspaper see Parker (2014).
53 Interview with Lorna Reid.
54 Interview with Grace Livingstone.
55 Interview with Liz Denver.
56 Interview with Mark Farmaner.
57 Interview with Cat Wiener.
58 Interview with David Yaffe.
59 Interview with Adam Bowles.

Bibliography

Archival material

Anti-Apartheid Movement Archive, Bodleian Library,

MSS AAM 21: *The Anti-Apartheid Movement and City AA: a statement by the AAM executive committee,* 1 December 1985.

MSS AAM 503: *Statement issued by the Anti-Apartheid Movement Executive Committee*, 10 July 1984.

British Library National Sound Archive

Kempster, J. (1988), 'Report on the Non-Stop Picket of South Africa House' (excerpt from *Black Londoners*, 18 April), *BBC Radio London*. C1499/1 C2.

British Universities Film and Video Council

LBC/IRN, "Kitson family on aunt's murder in South Africa" 15 January 1982. http://bufvc.ac.uk/tvandradio/lbc/index.php/segment/0000300173012 (Accessed 02 August 2016).

LBC/IRN, "Norma Kitson on South Africa", September 1986. http://bufvc.ac.uk/tvandradio/lbc/index.php/segment/0011900056007 (Accessed 02 August 2016).

City of London Anti-Apartheid Group papers

City of London Anti-Apartheid Group (1984), 'Press statement – 10 June 1984'.

City of London Anti-Apartheid Group, Non-Stop Picket stewards' books, an incomplete set covering most of the period 19 April 1986 – 23 February 1990.

City of London Anti-Apartheid Group (1986) *100 Days and Nights: A record of police harassment*, 28 July.

City of London Anti-Apartheid Group (1986) *Non-Stop News*, 9, 8 August.

City of London Anti-Apartheid Group (1986) *Non-Stop News*, 11, 19 September.

City of London Anti-Apartheid Group (1986) *Non-Stop News*, 12, 10 October.

City of London Anti-Apartheid Group (1986) 'Non-Stop Picket until Mandela is Free,' A5 leaflet, November/December.

City of London Anti-Apartheid Group (1987) *Non-Stop News*, 16, 23 January.

Bibliography

City of London Anti-Apartheid Group (1987) *Non-Stop News*, Anniversary Issue, April.

City of London Anti-Apartheid Group (1987) *Non-Stop Against Apartheid*, 23, October.

City of London Anti-Apartheid Group (1987) *Non-Stop Against Apartheid*, 24, December.

City of London Anti-Apartheid Group (1988) *Non-Stop Against Apartheid*, 25, January/February.

City of London Anti-Apartheid Group (1988) *Picketers News*, 21–26 January.

City of London Anti-Apartheid Group (1988) *Non-Stop Against Apartheid*, 26, March.

City of London Anti-Apartheid Group (1988) *Picketer's News*, 20 March.

City of London Anti-Apartheid Group (1988) *Non-Stop Against Apartheid*, 27, April.

City of London Anti-Apartheid Group (1988) *Picketer's News*, 27 May.

City of London Anti-Apartheid Group (1988) *Non-Stop Against Apartheid*, 28, June.

City of London Anti-Apartheid Group (1988) *Picketers News*, 10 June.

City of London Anti-Apartheid Group (1988) 'No Rights? No Flights!' A5 leaflet, July.

City of London Anti-Apartheid Group (1988) *Non-Stop Against Apartheid*, 29, July.

City of London Anti-Apartheid Group (1988) 'Office diary 2 July'.

City of London Anti-Apartheid Group (1988) *Picketer's News*, 8 July.

City of London Anti-Apartheid Group (1988) *Picketer's News*, 22 July.

City of London Anti-Apartheid Group (1988) *Picketer's News*, 12 August.

City of London Anti-Apartheid Group (1989) 'Surround the Racist Embassy,' A5 leaflet, May/June.

City of London Anti-Apartheid Group (1989) *Non-Stop Against Apartheid*, 34, June.

City of London Anti-Apartheid Group (1989) *Non-Stop Against Apartheid*, 35, July/August.

City of London Anti-Apartheid Group (1989) *Annual Report 1988–89: 365 Days in Struggle*, 23 September.

City of London Anti-Apartheid Group (1989) *Non-Stop Against Apartheid*, 36, October/November.

City of London Anti-Apartheid Group (1990) *The Non-Stop Picket: A user friendly guide*, unpublished internal document.

City of London Anti-Apartheid Group (1990) *Non-Stop Against Apartheid*, Special Issue (38), February.

City of London Anti-Apartheid Group (1990), Press statement about the arrest of 11 City of London Ant-Apartheid Group protestors against Safeway's, in Walworth Road, stocking South African fruit and vegetables – in solidarity with Zebediela strikers, 17 August.

Farmaner, M. (1990), *Record of Police Harassment on and around the NSP, 28th September 1989-4th January 1990*, Internal report for the City of London Anti-Apartheid Group Committee.

Handwritten 'report on the brazier' by Richard Roques presented to the City Group committee, 26 January 1987.

Higginbottom, A. (1986) 'Report on Defence Campaigns', paper submitted to the City of London Anti-Apartheid Group Annual General Meeting, held at County Hall, London, 16 February 1986.

Justice for Kitson Campaign (1988) *Justice for Kitson Campaign*, A5 pamphlet.

Bibliography 223

Letter from Andy Higginbottom to Councillor Andrew Dismore, Westminster City Council, 18 October 1986; letter from Andrew Dismore to Andy Higginbottom, 26 October 1986; letter on behalf of Andrew Dismore to City Group, 11 December 1986.

Letter from Andy Higginbottom to Councillor Bob Crossman, Mayor of London Borough of Islington, 8 December 1986.

Letter from City Group to members and supporters titled 'Forthcoming release of Nelson Mandela', 18 January 1990.

Letter from Committee for the Defence of Democratic Rights in Turkey inviting participation in their event, 27 October 1986.

Letter from Don Nkadimeng, Secretary General of AZAPO, to City Group, 6 October 1992.

Letter from Free the Guildford Four campaign inviting City Group to sponsor their campaign, 25 June 1986.

Letter from Lambeth Women's Rights Committee to invite City Group Singers to perform at 'Women Internationally' day of activities, 3 March 1987.

Letter from Mpotseng Jairus Kgokong, Secretary General of the Black Consciousness Movement (of Azania), to City Group, 29 September 1992.

Letter from Ravinder Bhogal, the black workers' group coordinator at Brent Council, to City Group, 29 September 1987.

Letter from Richard Roques, on behalf of City Group, to Bishop Trevor Huddleston, 25 May 1987.

Letter from Richard Roques, on behalf of the Springbok Reception Committee, to the London ANC, 27 October 1992.

Letter from SACOS Rugby Union to City Group, 1992 (undated).

Letter from Simon Murray to City Group Committee, 21 August 1989.

Minutes of a meeting of 'non-aligned' activists held at Norma Kitson's home, 5 July 1986.

Witness statement of Councillor A in relation to arrest at South African Airways, 2 July 1988.

Witness statement of Councillor B in relation to arrest at South African Airways, 2 July 1988.

Witness statement of Kathleen Fernand in relation to arrest at South African Airways, 2 July 1988.

Independent Television News Source (online archive)

South Africa, 19 October 1985: www.itnsource.com/en/shotlist/ITN/1985/10/19/AS191085008/ (Accessed 24 April 2017).

South Africa, Mandela Release Special, 11 February 1990: www.itnsource.com/en/shotlist/ITN/1990/02/11/BSP110290007/ (Accessed 24 April 2017).

UK: 12-year-old Steven Kitson in 1969 anti-apartheid rally; Labour politician Tony Benn among 1982 protestors seeking release of Steven and his father' (2 clips): www.itnsource.com/en/shotlist/RTV/1982/01/11/RTV110182002/ (Accessed 24 April 2017).

Mayibuye Archive, University of Western Cape

MCH-02 ANC London Box 14, F, untitled bright blue filing cabinet folder: 'Memo from the RPC further to the recommendation for the expulsion from membership of the ANC of Norma and David Kitson', 3 February 1987.

224   Bibliography

Steven Kitson's papers

South African Embassy Picket Campaign (SAEPC) folder, 1984: 'London Borough of Camden powers of demonstration', legal opinion by Stephen Sedley QC, prepared for the London Borough of Camden, 6 July 1984.

South African Embassy Picket Campaign (SAEPC) folder, 1984: Witness Statement by Commander George Howlett of Cannon Row Police Station, Metropolitan Police, 6 July 1984.

Interviews

Email correspondence with Andrew Crawley, 6 December 2012.
Email correspondence with anonymous retired Chief Superintendent, 11 December 2012.
Email correspondence with anonymous retired police constable, 5 December 2012.
Email correspondence with anonymous retired woman police constable, 10 December 2012.
Email correspondence with Chris, 5 December 2013.
Email correspondence with David Press, 16 December 2013.
Email correspondence with Debbie, 7 December 2013.
Email correspondence with Fiona Brownlie, 13 January 2013.
Email correspondence with former Inspector David Lee, 4 December 2012.
Email correspondence with Graham Neale, 6 December 2013.
Email correspondence with Jo Cooper, 19 February 2014.
Email correspondence with Julie, 13 January 2014.
Email correspondence with Keith Veness, 13 January 2014.
Email correspondence with Liz Myers, 13 January 2014.
Email correspondence with Lorna Whitfield, 16 December 2013.
Email correspondence with Mark Bearn, 13 January 2014.
Email correspondence with Mark Montgomerie, 13 January 2014.
Email correspondence with Nick Westwood, 4 December 2012.
Email correspondence with retired Deputy Assistant Commissioner David Gilbertson, 5 September 2013.
Email correspondence with Sally O'Donnell, 2 January 2014.
Interview with Adam Bowles, 28 November 2013.
Interview with Amanda Collins, 16 January 2014.
Interview with Andre Schott, 9 April 2013.
Interview with Andy Higginbottom, 17 April 2012.
Interview with Andy Privett, 12 March 2012.
Interview with Ann Elliot, 29 May 2013.
Interview with anonymous retired Woman Police Constable from Special Patrol Group, February 2013.
Interview with April, 12 April 2013.
Interview with Ben Evans, 15 August 2013.
Interview with Betta Garavaldi, 1 November 2011.
Interview with Bob Myers, 23 April 2014.
Interview with Carol Brickley, 21 February 2013.
Interview with Cat Weiner, 5 March 2013.
Interview with Charine John, 4 June 2013.

Interview with Chris Proctor, 2 October 2013.
Interview with Claus, 5 October 2011.
Interview with Daniel Jewesbury, 28 June 2013.
Interview with Danny Simpson, 26 February 2014.
Interview with David Yaffe, 5 April 2013
Interview with Deborah Potts, 8 May 2013.
Interview with Deirdre Healy, 10 March 2013.
Interview with Dominic Thackray, 3 April 2013.
Interview with Eleanor Cave, 20 September 2013.
Interview with Eric Levy, 16 April 2013.
Interview with Francis Squire, 25 November 2011.
Interview with Gary Lowe, 3 May 2013.
Interview with Georgina Lansbury, 30 July 2013.
Interview with Gerald Denver, 22 April 2013.
Interview with Grace Livingstone, 2 August 2013.
Interview with Hannah Caller, 19 June 2013
Interview with Helen Landau, 13 December 2011.
Interview with Helen Marsden, 21 August 2013.
Interview with Hema Patel, 6 August 2013.
Interview with Irene Minczer, 19 July 2013.
Interview with Jacky Sutton, 23 August 2011.
Interview with Janet Boateng, 7 February 2014.
Interview with James Godfrey 30 January 2013.
Interview with Jeremy Corbyn MP, 6 September 2013.
Interview with Kathy Fernand, 8 July 2013.
Interview with Ken Bodden, 6 March 2013.
Interview with Liz Denver, 22 April 2013.
Interview with Lorna Reid, 16 May 2013.
Interview with Louise Christian, 2 December 2013.
Interview with Lucie Smoker, 20 March 2012.
Interview with Lynne Reid Banks, 9 December 2013.
Interview with Maggie Mellon, 13 June 2013.
Interview with Mark Farmaner, 10 April 2013.
Interview with Mike Burgess, 28 January 2013.
Interview with Nick Manley, 28 November 2012.
Interview with Nicki, 27 March 2013.
Interview with Nicole Aebi, 28 November 2011.
Interview with Patrick C, 13 January 2013.
Interview with Paul Boateng, 7 February 2014.
Interview with Peter Tatchell, 19 December 2013.
Interview with Penelope Reynolds, 12 July 2013.
Interview with Razia Meer, 26 September 2013.
Interview with Rebecca Copas, 9 September 2011
Interview with Richard Roques, 22 March 2013.
Interview with Rikke Nielsen, 5 October 2011.
Interview with Ross Galbraith, 4 March 2013.
Interview with Ruby Noorani, 21 June 2013.
Interview with Sam Smith, 14 August 2013.
Interview with Selman Ansari, 14 May 2013.

Interview with Shafeeq Meer, 26 September 2013.
Interview with Sharon Chisholm, 13 September 2013.
Interview with Shereen Pandit, 31 May 2013.
Interview with Sian Newman, 21 October 2011.
Interview with Sigþrúður Gunnarsdóttir, 23 November 2011.
Interview with Simon Murray, 5 March 2013.
Interview with Simone Maloney, 7 April 2013.
Interview with Sinead, 9 March 2013.
Interview with Susan Yaffe, 31 May 2013.
Interview with Thabo, 9 July 2013.
Interview with Trevor Rayne, 21 March 2014.
Interview with Tunde Forrest, 26 April 2013.
Interview with Vincent, 1 December 2012.

Published work

Abercrombie, N., Warde, A., Soothill, K., Urry, J. and Walby, S. (1988) *Contemporary British Society. A New Introduction to Sociology*, Cambridge: Polity Press.
Alexander, P. (2013) 'Marikana, turning point in South African history,' *Review of African Political Economy* 40(138): 605–619.
Amin, A. and Graham, S. (1997) 'The ordinary city,' *Transactions of the Institute of British Geographers* 22: 411–429.
Anderson, B. (2004) 'Time-stilled space-slowed: how boredom matters,' *Geoforum* 35(6): 739–754.
Anderson, B. (2014) *Encountering Affect: Capacities, Apparatuses, Conditions*, Farnham: Ashgate.
Askins, K. (2015) 'Being together: everyday geographies and the quiet politics of belonging,' *ACME: An International E-journal for Critical Geographies* 14(2): 470–478.
Bailey, S. and Taylor, N. (2009) *Civil Liberties Cases, Materials and Commentary*, 6th edition, Oxford: Oxford University Press.
Ball, J. (1994) *The Ritual of the Necklace*. Research report written for the Centre for the Study of Violence and Reconciliation, March. Available online at www.csvr.org.za/index.php/publications/1632-the-ritual-of-the-necklace.html (last accessed 17 April 2017).
Bell, T. (2009) *Comrade Moss. A Political Journey*, Muizenberg, SA: RedWorks.
Bell, T. with Ntsebeza, D.B. (2003) *Unfinished Business: South Africa, Apartheid and Truth*, London: Verso.
Bellamy, R. (1994) 'The anti-poll tax non-payment campaign and liberal concepts of political obligation,' *Government and Opposition* 29(1): 22–41.
Benn, T. (1994) *The End of an Era. Diaries 1980–90*, London: Arrow Books.
Benwell, M. and Hopkins, P. (eds.) (2016) *Children, Young People and Critical Geopolitics*, Farnham: Ashgate.
Benyon, J. and Solomos, J. (1988) 'The simmering cities: urban unrest during the Thatcher years,' *Parliamentary Affairs* 41(3): 402–422.
Biko, S. (2015) *I Write What I Like: Selected Writings*, Chicago, IL: University of Chicago Press.
Bissell, D. (2007) 'Animating suspension: waiting for mobilities,' *Mobilities* 2(2): 277–298.
Bissell, D. (2015) 'Virtual infrastructures of habit: the changing intensities of habit through gracefulness, restlessness and clumsiness,' *cultural geographies* 22(1): 127–146.

Bond, P. (2000) *Elite Transitions: From Apartheid to Neoliberalism in South Africa*, London: Pluto.
Booth, D. (2003) 'Hitting apartheid for six? The politics of the South African sports boycott,' *Journal of Contemporary History* 38(3): 477–493.
Bosco, F.J. (2006) 'The Madres de Plaza de Mayo and three decades of human rights' activism: embeddedness, emotions, and social movements,' *Annals of the Association of American Geographers* 96(2): 342–365.
Bosco, F.J. (2007) 'Emotions that build networks: geographies of human rights movements in Argentina and beyond,' *Tijdschrift voor Econmomische en Sociale Geografie* 98(5): 545–563.
Bosco, F.J. (2010) 'Play, work or activism? Broadening the conections between the political and children's geographies,' *Children's Geographies* 8(4): 381–390.
Bourdieu, P. (1977) *Outline of a Theory of Practice*, Cambridge: Cambridge University Press.
Bourdieu, P. (1984) *Distinction: A Social Critique of the Judgement of Taste*, Cambridge, MA: Harvard University Press.
Bowlby, S. (2011) 'Friendship, co-presence and care: neglected spaces,' *Social & Cultural Geography* 12(6): 605–622.
Bozzoli, B. (2004) *Theatres of Struggle and the End of Apartheid*, Athens: Ohio University Press.
Bozzoli, B. with Nkotsoe, M. (1991) *Women of Phokeng. Consciousness, Life Strategy, and Migrancy in South Africa, 1900–1983*, London: James Currey.
Brickley, C. (1985) 'South Africa and the global capitalist economy,' *Contemporary Issues in Geography and Education* 2(1): 8–11.
Brickley, C., O'Halloran, T. and Reed, D. (1986) *South Africa: Britain Out of Apartheid, Apartheid Out of Britain*, 2nd edition, London: Larkin Publications.
Brown, G. (2008) 'Ceramics, clothing and other bodies: affective geographies of homoerotic cruising encounters,' *Social and Cultural Geography* 9(8): 915–932.
Brown, G. (2009) 'Thinking beyond homonormativity: performative explorations of diverse gay economies,' *Environment & Planning A* 41: 1496–1510.
Brown, G. (2013) 'Unruly bodies (standing against apartheid),' in A. Cameron, J. Dickinson and N. Smith (eds.), *Body-States*, Aldershot: Ashgate, pp. 145–157.
Brown, G. (2018) 'Anti-apartheid solidarity in the perspectives and practices of the British far left in the 1970s and '80s,' in E. Smith and M. Worley (eds.), *Waiting for the Revolution: The British Far Left from 1956 vol. II.*, Manchester: Manchester University Press, pp. 66–87.
Brown, G. and Pickerill, J. (2009) 'Space for emotion in the spaces of activism,' *Emotion, Space and Society* 2(1): 24–35.
Brown, G. and Yaffe, H. (2013) 'Non-Stop against apartheid: practicing solidarity outside the South African embassy,' *Social Movement Studies* 12(2): 227–234.
Brown, G. and Yaffe, H. (2014) 'Practices of solidarity: opposing apartheid in the centre of London,' *Antipode* 46(1): 34–52.
Brown, G. and Yaffe, H. (2016) 'Young people's engagement with the geopolitics of anti-apartheid solidarity in 1980s' London,' in M. Benwell and P. Hopkins (eds.), *Children, Young People and Critical Geopolitics*, Farnham: Ashgate, pp. 155–168.
Brown, G., Feigenbaum, A., Frenzel, F. and McCurdy, P. (eds.) (2017) *Protest Camps in International Context: Spaces, Infrastructures, and Media of Resistance*, Bristol: Policy Press.
Buck, N., Gordon, I., Hall, P., Harloe, M. and Kleinman, M. (2002) *Working Capital: Life and Labour in Contemporary London*, London: Routledge.

Buechler, S.M. (2011) *Understanding Social Movements: Theories from the Classical Era to the Present*, Boulder, CO: Paradigm Publishers.
Bunce, R. and Field, P. (2014) *Darcus Howe. A Political Biography*, London: Bloomsbury.
Bundy, C. (1989) 'Around which corner? Revolutionary theory and contemporary South Africa,' *Transformation* 8: 1–23.
Bunnell, T., Yea, S., Peake, L., Skelton, T. and Smith, M. (2012) 'Geographies of friendships,' *Progress in Human Geography* 36(4): 490–507.
Burridge, A. (2010) 'Youth on the line and the No Borders movement,' *Children's Geographies* 8(4): 401–411.
Burton-Cartledge, P. (2014) 'Marching separately, seldom together: The political history of two principal trends in British Trotskyism, 1945–2009,' in E. Smith and M. Worley (eds.), *Against the Grain: The British Far Left from 1956*, Manchester: Manchester University Press, pp. 80–97.
Butler, J., Gambetti, Z. and Sabsay, L. (eds.) (2016) *Vulnerability in Resistance*, Durham, NC: Duke University Press.
Butler, T. and Rustin, M. (eds.) (1996) *Rising in the East. The Regeneration of East London*, London: Lawrence & Wishart.
Byrd, P. (ed.) (1988) *British Foreign Policy under Thatcher*, Oxford: Philip Allan.
Chari, S. and Donner, H. (2010) 'Ethnographies of Activism: A Critical Introduction,' *Cultural Dynamics* 22(2): 75–85.
Chatterton, P. (2006) "Give up activism' and change the world in unknown ways: or, learning to walk with others on uncommon ground,' *Antipode* 38(2): 259–281.
Chatterton, P., Featherstone, D., & Routledge, P. (2013) 'Articulating climate justice in Copenhagen: antagonism, the commons, and solidarity,' *Antipode* 45(3): 602–620.
Christian, L. (1987) "Instructions from on high' – Policing the Picket outside the South African Embassy,' *Socialist Lawyer* 3 (Autumn): 15–17. Available online: https://static1.squarespace.com/static/562e7d33e4b0da14ad6d202f/t/566dcb054bf118e6b44fb2f0/1450035973936/SocialistLawyer03.pdf (last accessed 7 January 2017).
Centre for Contemporary Cultural Studies (1982) *The Empire Strikes Back: Race and Racism in 70s Britain*, London: Hutchinson Educational.
Cohen, S.B. (2015) *Geopolitics. The geography of international relations*, 3rd edition, London: Rowman & Littlefield.
Conlon, D. (2011) 'Waiting: feminist perspectives on the spacings/timings of migrant (im)mobility,' *Gender, Place and Culture* 18(3): 353–360.
Cox, L. (2010) 'How do we keep going? Activist burnout and personal sustainability in social movements,' paper presented at *Fourteenth international conference on alternative futures and popular protest*, Manchester Metropolitan University. Available online at: http://eprints.maynoothuniversity.ie/2201/1/LC-How_do_we_keep_going.pdf (last accessed 14 August 2016).
Craggs, R. (2014a) 'Hospitality in geopolitics and the making of Commonwealth international relations,' *Geoforum* 52: 90–100.
Craggs, R. (2014b) 'Postcolonial geographies, decolonization, and the performance of geopolitics at Commonwealth conferences,' *Singapore Journal of Tropical Geography* 35(1): 39–55.
Craggs, R. and Mahony, M. (2014) 'The geographies of the conference: knowledge, performance and protest,' *Geography Compass* 8(6): 414–430.
Cresswell, T. (1994) 'Putting women in their place: the carnival at Greenham Common,' *Antipode* 26(1): 35–58.

Cumbers, A. and Routledge, P. (2013) 'Place, space and solidarity in global justice networks,' in D. Featherstone and J. Painter (eds.) *Spatial Politics. Essays for Doreen Massey*, Oxford: Wiley Blackwell, pp. 213–223.

Cumbers, A., Routledge, P. and Nativel, C. (2008) 'The entangled geographies of global justice networks,' *Progress in Human Geography* 32(2): 183–201.

Darling, J. (2009) 'Thinking beyond place: the responsibilities of a relational spatial politics,' *Geography Compass* 3(5): 1938–1954.

Davies, A. (2012) 'Assemblage and social movements: Tibet support groups and the spatialities of political organisation,' *Transactions of the Institute of British Geographers* 37(2): 273–286.

De Gruchy, J.W. (1986) 'The church and the struggle for South Africa,' *Theology Today* 43(2): 229–243.

De Jager, N. with Graham, V., Gumede, V., Mangcu, X., Neethling, T., Steyn Kotze, J. and Welsh, D. (2015) *South African Politics. An Introduction*, Cape Town: Oxford University Press (Southern Africa).

Della Porta, D. and Diani, M. (1999) *Social Movements: An Introduction*, Oxford: Blackwell Publishing.

Dewsbury, J.D. and Bissell, D. (2015) 'Habit geographies: the perilous zones in the life of the individual,' *cultural geographies* 22(1): 21–28.

Dickens, L. and MacDonald, R.L. (2015) '"I can do things here that I can't do in my own life": the making of a civic archive at the Salford Lads Club,' *ACME: An International E-Journal for Critical Geographies* 14(2): 377–389.

Dittmer, J. (2013) 'Humour at the Model United Nations: the role of laughter in constituting geopolitical assemblages,' *Geopolitics* 18(3): 493–513.

Dittmer, J. (2014) 'Geopolitical assemblages and complexity,' *Progress in Human Geography* 38(3): 385–401.

Dittmer, J. and McConnell, F. (eds.) (2015) *Diplomatic Cultures and International Politics: Translations, Spaces and Alternatives*, London: Routledge.

Dodds, K. (2005) *Global Geopolitics: A Critical Introduction*, Harlow: Pearson Education.

Dodds, K. and Kirby, P. (2013) 'It's not a laughing matter: critical geopolitics, humour and unlaughter,' *Geopolitics* 18(1): 45–49.

Durbach, A. (1999) *Upington: A Story of Trials and Reconciliation*, St Leonards, NSW: Allen & Unwin.

Edensor, T. (ed.) (2012) *Geographies of Rhythm: Nature, Place, Mobilities and Bodies*, Aldershot: Ashgate.

Ellis, C. (2004) *The Ethnographic I: A Methodological Novel about Autoethnography*, Oxford: AltaMira Press.

Ellis, S. (2013) *External Mission: The ANC in Exile 1960–1990*, London: Hurst & Co.

Ellis, S. and Sechaba, T. (1992) *Comrades Against Apartheid: The ANC and the South African Communist Party in Exile*, Bloomington, IN: Indiana University Press.

Evans, E. (2013) *Thatcher and Thatcherism*, 3rd edition, London: Routledge.

Evans, R. and Lewis, P. (2013) *Undercover: The True Story of Britain's Secret Police*, London: Faber and Faber.

Featherstone, D. (2008) *Resistance, Space and Political Identities*, Oxford: Wiley-Blackwell

Featherstone, D. (2012) *Solidarity: Hidden Histories and Geographies of Internationalism*, London: Zed.

Featherstone, D. and Painter, J. (eds.) (2013) *Spatial Politics. Essays for Doreen Massey*, Oxford: Wiley Blackwell.

Feigenbaum, A. (2013) 'Written in the mud: (proto) zine-making and autonomous media at the Greenham Common Women's Peace Camp,' *Feminist Media Studies* 13(1): 1–13.

Feigenbaum, A., Frenzel, F. and McCurdy, P. (2013) *Protest Camps*, London: Zed.

Feinstein, A. (2009) *After the Party: Corruption, the ANC and South Africa's Uncertain Future*, London: Verso.

Fieldhouse, R. (2005) *Anti-Apartheid. A History of the Movement in Britain*, London: Merlin Press.

Fine, R. (1989) 'The antinomies of nationalism and democracy in the South African liberation struggle,' *Review of African Political Economy* 45/46: 98–106.

Flood, C. and Grindon, G. (eds.) (2014) *Disobedient Objects*, London: V&A Publishing.

Franks, B. (2006) *Rebel Alliances: The Means and Ends of Contemporary British Anarchisms*, Edinburgh: AK Press.

Gaskell, C. (2008) "But they just don't respect us': young people's experiences of (dis)respected citizenship and the New Labour Respect Agenda,' *Children's Geographies* 6(3): 223–238.

Gibson, N.C. (2011) *Fanonian Practices in South Africa: From Steve Biko to Abahlali BaseMjondolo*, Basingstoke: Palgrave Macmillan.

Gibson, N.C. (2012) 'What happened to the "promised land"? A Fanonian perspective on post-apartheid South Africa,' *Antipode* 44(1): 51–73.

Giddens, A. (1991) *Modernity and Self-identity: Self and Society in the Late Modern Age*, Stanford, CA: Stanford University Press.

Gilbert, S. (2007) 'Singing against apartheid: ANC cultural groups and the international anti-apartheid struggle,' *Journal of Southern African Studies* 33(2): 421–441.

Gilroy, P. (2004) *After Empire: Melancholia or Convivial Culture?* London: Routledge.

Goodwin, J., Jasper, J.M. and Polletta, F. (eds.) (2001) *Passionate Politics: Emotions and Social Movements*, Chicago, IL: University of Chicago Press.

Gordon, H.R. (2008) 'Gendered paths to teenage political participation: parental power, civic mobility, and youth activism,' *Gender & Society* 22(1): 31–55.

Gould, C.C. (2007), 'Transnational solidarities,' *Journal of Social Philosophy* 38(1): 148–164.

Graham, S. (2010) 'When infrastructures fail,' in S. Graham (ed.) *Disrupted Cities: When Infrastructures Fail*, London: Routledge, pp. 1–26.

Guelke, A. (2000) 'Ireland and South Africa: a very special relationship,' *Irish Studies in International Affairs* 11: 137–146.

Guelke, A. (2005) *Rethinking the Rise and Fall of Apartheid*, Basingstoke: Palgrave MacMillan.

Gurney, C. (2000) '"A great cause": the origins of the anti-apartheid movement, June 1959-March 1960,' *Journal of Southern African Studies* 26(1):123–144.

Halvorsen, S. (2015) 'Taking space: moments of rupture and everyday life in occupy London,' *Antipode* 47(2): 401–417.

Halvorsen, S (2017) 'Losing space in occupy London: fetishising the protest camp,' in G. Brown, A. Feigenbaum, F. Frenzel and P. McCurdy (eds.) *Protest Camps in International Context: Spaces, Infrastructures and Media of Resistance*, Bristol: Policy Press, pp. 163–178.

Harris, C. and Valentine, G. (2016) 'Childhood narratives: adult reflections on encounters with difference in everyday spaces,' *Children's Geographies*, doi: 10/1080/14733285.2016.1269153.

Hart, M. (2007) 'Humour and social protest: an introduction,' *International Review of Social History* 52: 1–20.
Harvey, D. and Williams, R. (1995) 'Militant particularism and global ambition: the conceptual politics of place, space, and environment in the work of Raymond Williams,' *Social Text* 42: 69–98.
Hayes, M. (2014) 'Red Action – left-wing political pariah. Some observations regarding ideological apostasy and the discourse of proletarian resistance,' in E. Smith and M. Worley (eds.) *Against the Grain: The British Far Left from 1956*, Manchester: Manchester University Press, pp. 229–246.
Herbstein, D. (2004) *White Lies: Canon Collins and the Secret War Against Apartheid*, Woodbridge: James Currey Publishers.
Hickman, M.J. (1998) 'Reconstructing deconstructing 'race': British political discourses about the Irish in Britain,' *Ethnic and Racial Studies* 21(2): 288–307.
Hillyard, P. (1993) *Suspect Community: People's Experience of the Prevention of Terrorism Acts in Britain*, London: Pluto Press (in association with the National Council for Civil Liberties).
Hitchings, R. (2012) 'People can talk about their practices,' *Area* 44(1): 61–67.
Hoffman, E. (2011) *Time*, London: Profile Books.
Høgsbjerg, C. (2014) *CLR James in Imperial Britain*, Durham, NC: Duke University Press.
Hopkins, P. (2010) *Young People, Place and Identity*, London: Routledge.
Hopkins, P. and Pain, R. (2007) 'Geographies of age: thinking relationally,' *Area* 39(3): 287–294.
Hörschelmann, K (2008) 'Populating the landscapes of critical geopolitics – young people's responses to the war on Iraq,' *Political Geography* 27(5): 587–609.
Hörschelmann, K. (2011) 'Theorising life transitions: geographical perspectives,' *Area* 43(4): 378–383.
Horton, J. and Kraftl, P. (2006) 'Not just growing up, but going on: materials, spacings, bodies, situations,' *Children's Geographies* 4(3): 259–276.
Hubbard, P. (2013) *Cities and Sexualities*, London: Routledge.
Israel, M. (1998) 'Crimes of the state: victimisation of South African political exiles in the United Kingdom,' *Crime, Law and Social Change* 29(1): 1–29.
Iveson, K. (2017) "Making space public' through occupation: the Aboriginal Tent Embassy, Canberra,' *Environment and Planning A* 49(3): 537–554
Jackson, P. (2004) 'Local consumption cultures in a globalizing world,' *Transactions of the Institute of British Geographers* 29(2): 165–178.
Jackson, B. and Saunders, R. (eds.) (2012) *Making Thatcher's Britain*, Cambridge: Cambridge University Press.
Jacobs, J. (1996) *Edge of Empire: Postcolonialism and the City*, London: Routledge.
Jasper, J. (1997) *The Art of Moral Protest: Culture, Biography and Creativity in Social Movements*, Chicago, IL: University of Chicago Press.
Jaster, R.S. (1990) *The 1988 Peace Accords and the Future of South-Western Africa*, Oxford: Brassey's for the International Institute for Strategic Studies.
Jeenah, N. (2015) 'Jihad as a form of struggle in the resistance to apartheid in South Africa,' in E. Kendall and E. Stein (eds.) *Twenty-first Century Jihad: Law, Society and Military Action*, London: IB Tauris, pp. 201–215.
Jeffrey, A. (2012) *The Improvised State: Sovereignty, Performance and Agency in Dayton Bosnia*, Oxford: Wiley.
Jeffrey, C. (2008) 'Waiting,' *Environment and Planning D: Society and Space* 26: 954–958.

Jeffrey, C. (2010) *Timepass: Youth, Class and the Politics of Waiting in India*, Palo Alto, CA: Stanford University Press.
Jeffrey, C. (2013) 'Geographies of children and youth III: alchemists of the revolution?' *Progress in Human Geography* 37(1): 145–152.
Johnsen, S., Cloke, P. and May, J. (2005) 'Transitory spaces of care: serving homeless people on the street,' *Health & Place* 11(4): 323–336.
Juris, J.S. and Pleyers, G.H. (2009) 'Alter-activism: emerging cultures of participation among young global justice activists,' *Journal of Youth Studies* 12(1): 57–75.
Kallio, K.P. and Hakli, J. (2011) 'Tracing children's politics,' *Political Geography* 30: 99–109.
Kanngieser, A. (2012) 'A sonic geography of voice towards an affective politics,' *Progress in Human Geography* 36(3): 336–353.
Katz, C. (2001) 'On the grounds of globalization: a topography for feminist political engagement,' *Signs* 26(4): 1213–1234.
Kay, J. (2010) *Red Dust Road: An Autobiographical Journey*, London: Picador.
Keable, K. (ed.) (2012) *The London Recruits*, London: The Merlin Press.
Keith, M. (2005) *After the Cosmopolitan? Multicultural Cities and the Future of Racism*, London: Routledge.
Kelliher, D. (2014) 'Solidarity and Sexuality: Lesbians and Gays Support the Miners 1984–5,' *History Workshop Journal* 77(1): 240–262.
Kelliher, D. (2017) 'Constructing a Culture of Solidarity: London and the British Coalfields in the London 1970s,' *Antipode* 49(1): 106–124.
King, A.D. (1990) *Global Cities: Post-imperialism and the Internationalization of London*, London: Routledge.
Kitson, D. (1991) 'Is the SACP really communist?' *Work in Progress* 73: 27–30.
Kitson, F. (1971) *Low Intensity Operations: Subversion, Insurgency, Peace-keeping*, London: Faber and Faber.
Kitson, N. (1987) *Where Sixpence Lives*, London: Hogarth Press.
Klein, G. (2009) 'The British anti-apartheid movement and political prisoner campaigns, 1973–1980,' *Journal of Southern African Studies* 35(2): 455–470.
Kondlo, K. (2009) *In the Twilight of the Revolutions. The Pan Africanist Congress of Azania (South Africa) 1959–1994*, Basel: Basler Africka Bibliographien.
Koopman, S. (2008) 'Imperialism within: can the master's tools bring down empire?' *ACME: An International E Journal for Critical Geographies* 7(2): 283–307.
Kraftl, P., Horton, J. and Tucker, F. (eds.) (2012) *Critical Geographies of Childhood and Youth: Policy and Practice*, Bristol: Policy Press.
Kyle, R.G., Milligan, C., Kearns, R.A., Larner, W., Fyfe, N.R. and Bondi, L. (2011) 'The tertiary turn: locating 'the academy; in autobiographical accounts of activism in Manchester, UK and Auckland, Aotearoa New Zealand,' *Antipode* 43: 1181–1214.
Latour, B. (2005) *Reassembling the Social. An Introduction to Actor-Network-Theory*, Oxford: Oxford University Press.
Lee, R. (2009) *African Women and Apartheid: Migration and Settlement in South Africa*, London: IB Tauris.
Lefebvre, H. (2004) *Rhythmanalysis: Space, Time and Everyday Life* (trans. S. Elden and G. Moore), London: Continuum.
Leitner, H., Sheppard, E. and Sziarto, K.M. (2008) 'The spatialities of contentious politics,' *Transactions of the Institute of British Geographers* 33(2): 157–172.
Lester, A. (1998) *From Colonization to Democracy. A New Historical Geography of South Africa*, London: IB Tauris.

Lodge, T. (2006) "An'boks amach': the Irish Anti-Apartheid Movement,' *History Ireland* 14(4): 35–39.
Lodge, T. (2011) *Sharpeville: An Apartheid Massacre and Its Consequences*, Oxford: Oxford University Press.
Longhurst, R. (2001) *Bodies: Exploring Fluid Boundaries*, London: Routledge.
López, M.A.M. (2013) 'The squatters' movement in Europe: a durable struggle for social autonomy in urban politics,' *Antipode* 45(4): 866–887.
Maaba, B.B. (2001) 'The archives of the Pan-Africanist Congress and the Black Consciousness-orientated movements,' *History in Africa* 28: 417–438.
MacLean, M. (2010) 'Anti-apartheid boycotts and the affective economies of struggle: the case of Aotearoa New Zealand 1,' *Sport in Society* 13(1): 72–91.
Maddrell, A. (2016) 'Mapping grief. A conceptual framework for understanding the spatial dimensions of bereavement, mourning and remembrance,' *Social & Cultural Geography* 17(2): 166–188.
Maller, C.J. (2015) 'Understanding health through social practices: performance and materiality in everyday life,' *Sociology of Health & Illness*, 37(1): 52–66.
Maller, C. and Strengers, Y. (2013) 'The global migration of everyday life: investigating the practice memories of Australian migrants,' *Geoforum* 44: 243–252.
Mamadouh, V., Meijer, A., Sidaway, J.D. and van der Wusten, H. (2015) 'Toward an urban geography of diplomacy: lessons from The Hague,' *The Professional Geographer* 67(4): 564–574.
Mason, K. (2013) 'Academics and social movements: Knowing our place, making our space,' *ACME: An International E - Journal for Critical Geographies* 12(1): 23–43.
Massey, D. (2005) *For Space*, London: Sage.
Massey, D. (2007) *World City*, Cambridge: Polity.
Massey, D. (2008) 'Geographies of solidarities,' in N. Clark, D. Massey, and P. Sarre (eds.), *Material Geographies: A World in the Making*, London: Sage & Open University Press, pp. 311–362.
Masten, A.S. (2001) 'Ordinary magic: Resilience processes in development,' *American Psychologist* 56(3): 227–238.
Matejskova, T. and Leitner, H. (2011) 'Urban encounters with difference: the contact hypothesis and immigrant integration projects in eastern Berlin,' *Social & Cultural Geography* 12(7): 717–741.
Matera, M. (2015) *Black London: The Imperial Metropolis and Decolonization in the Twentieth Century*, Oakland: University of California Press.
Mattern, M. (1998) *Acting in Concert: Music, Community, and Political Action*, New Brunswick, NJ: Rutgers University Press.
McConnell, F., Moreau, T. and Dittmer, J. (2012) 'Mimicking state diplomacy: the legitimizing strategies of unofficial diplomacies,' *Geoforum* 43(4): 804–814
McFarlane, C. and Vasudevan, A. (2014) 'Informal infrastructures,' in P. Adey, D. Bissell, K. Hannam, P. Merriman and M. Sheller (eds.) *Handbook of Mobilities*, London and New York: Routledge, pp. 256–264.
McKay, G. (2004) 'Subcultural innovations in the Campaign for Nuclear Disarmament,' *Peace Review* 16(4): 429–438.
McKinley, M. (1991) 'Of "alien influences": accounting and discounting for the international contacts of the Provisional Irish Republican Army,' *Journal of Conflict Studies* 11(3): 7–35.
McLeod, J. and Thomson, R. (2009) *Researching Social Change: Qualitative Approaches*, London: Sage Publications.

McSmith, A. (2011) *No Such Thing as Society. A History of Britain in the 1980s*, London: Constable.

Menendez, R. (2013) 'The Socialist Transformation of Venzuela: the geographical dimension of political strategy,' in D. Featherstone and J. Painter (eds.), *Spatial Politics. Essays for Doreen Massey*, Oxford: Wiley Blackwell, pp. 224–234.

Merrifield, A. (2013) *The Politics of the Encounter: Urban Theory and Protest under Planetary Urbanization*, Athens: University of Georgia Press.

Metz, S. (1986), 'The anti-apartheid movement and the populist instinct in American politics,' *Political Science Quarterly* 101(3): 379–395.

Miller, B.A. (2000) *Geography and Social Movements: Comparing Antinuclear Activism in the Boston Area*, Minneapolis: University of Minnesota Press.

Mills, S. (2011), 'Be prepared: communism and the politics of scouting in 1950s Britain,' *Contemporary British History* 25(3): 429–450.

Mills, S. (2012) 'Young ghosts: ethical and methodological issues of historical research in children's geographies,' *Children's Geographies* 10(3): 357–363.

Mills, S. (2015) 'Geographies of youth work, volunteering and employment: the Jewish Lads' Brigade and Club in post-war Manchester,' *Transactions of the Institute of British Geographers* 40(4): 523–535.

Mills, S. and Kraftl, P. (eds.) (2014) *Informal Education, Childhood and Youth: Geographies, Histories, Practices*, Basingstoke: Palgrave Macmillan.

Mitchell, D. (2003) *The Right to the City: Social Justice and the Fight for Public Space*, London: The Guilford Press.

Mitchell, D. and Staeheli, L.A. (2005) 'Permitting protest: parsing the fine geography of dissent in America,' *International Journal of Urban and Regional Research* 29(4): 796–813.

Mitchell, K. and Elwood, S. (2012) 'Mapping children's politics: the promise of articulation and the limits of nonrepresentational theory,' *Environment and Planning D: Society and space* 30(5): 788–804.

Mort, F. (1998) 'Cityscapes: consumption, masculinities and the mapping of London since 1950,' *Urban Studies*, 35 (5/6): 889–907.

Nauright, J. (1997) *Sport, Cultures, and Identities in South Africa*, London: Leicester University Press.

Nicolini, D. (2012) *Practice Theory, Work, & Organization: An Introduction*, Oxford: Oxford University Press.

O'Neill, S. (1989), 'Demo of a thousand days', *City Limits* (12–19 January), London: London Voice Ltd, p. 7 [out of print].

Oldman, J. (2004) 'Beyond bricks and mortar,' in J. Roche, S. Tucker, R. Thomson and R. Flynn (eds.), *Youth in Society*, 2nd edition, London: Sage, pp. 112–119.

Onslow, S. (ed.) (2009) *Cold War in Southern Africa: White Power, Black Liberation*, London: Routledge.

O Tuathail, G. (1999) 'The ethnic cleansing of a 'safe area': the fall of Srebrenica and the ethics of un-governability,' in J. Proctor and D. Smth (eds.), *Geography and Ethics*, London: Routledge, pp. 120–131.

Pain, R. (2004) 'Introduction: children at risk?' *Children's Geographies* 2(1): 65–67.

Pain, R. and Francis, P. (2004) 'Living with crime: spaces of risk for homeless young people,' *Children's Geographies* 2(1): 95–110.

Pain, R., Panelli, R., Kindon, S. and Little, J. (2010) 'Moments in everyday/distant geopolitics: young people's fears and hopes,' *Geoforum* 41: 972–982.

Pantzar, M. and Shove, E. (2010) 'Temporal rhythms as outcomes of social practices: a speculative discussion,' *Ethnologia Europaea* 40(1): 19–29.
Parker, L. (2014) 'Opposition in slow motion: the CPGB's 'anti-revisionists' in the 1960s and 1970s,' in E. Smith and M. Worley (eds.), *Against the Grain: The British Far Left from 1956*, Manchester: Manchester University Press, pp. 98–114.
Philo, C. and Smith, F. (2003) 'Guest editorial: political geographies of children and young people,' *Space and Polity* 7(2): 99–115.
Pilcher, J. and Wragg, S. (eds.) (1996) *Thatcher's Children? Politics, Childhood and Society in the 1980s and 1990s*, London: Falmer Press.
Pimlott, H. (2011) "Eternal ephemera' or the durability of 'disposable literature': the power and persistence of print in an electronic world,' *Media, Culture & Society* 33(4): 515–530.
Pogrund, B. (1990) *How Can Man Die Better... Sobukwe and Apartheid*, London: Peter Halban.
Pollock, J. (2004) "We don't want your racist tour': the 1981 Springbok tour and the anxiety of settlement in Aotearoa/New Zealand,' *Graduate Journal of Asia-Pacific Studies* 2(1): 32–43.
Polletta, F. (2012) *Freedom is an Endless Meeting: Democracy in American Social Movements*, Chicago, IL: University of Chicago Press.
Reckwitz, A. (2002) 'Toward a theory of social practices a development in culturalist theorizing,' *European Journal of Social Theory* 5(2): 243–263.
Reed, D. (1984) *Ireland, the Key to the British Revolution*, London: Larkin Publications. Available online at: www.revolutionarycommunist.org/images/pdf/IKBR_2016_IA.pdf (last accessed 3 April 2017).
Reid Banks, L. (1998) *Fair Exchange*, London: Piatkus.
Revill, G. (2016) 'How is space made in sound? Spatial mediation, critical phenomenology and the political agency of sound,' *Progress in Human Geography* 40(2): 240–256.
Revolutionary Communist Group. (1984) *The Revolutionary Road to Communism in Britain. Manifesto of the Revolutionary Communist Group*, London: Larkin Publications.
Reynolds, J.T. (2015) *Sovereignty and Struggle: Africa and Africans in the Era of the Cold War, 1945–1994*, Oxford: Oxford University Press.
Richardson, M. (2003) 'Leadership and mobilization: SOGAT in the 1986–87 News International dispute,' *Historical Studies in Industrial Relations* 15: 73–93.
Roberts, A. (1980) *The Rossing Files: The Inside Story of Britain's Secret Contract for Namibian Uranium*, London: Namibia Support Committee (CANUC).
Robinson, J. (2006) *Ordinary Cities: Between Modernity and Development*, London: Routledge.
Roseneil, S. (2000) *Common Women, Uncommon Practices: The Queer Feminisms of Greenham*, London: Cassell.
Routledge, P. (2009) 'Transnational resistance: Global justice networks and spaces of convergence,' *Geography Compass* 3(5): 1881–1901.
Routledge, P. (2010) 'Nineteen days in April: Urban protest and democracy in Nepal,' *Urban Studies* 47(6): 1279–1299.
Routledge, P. (2012) 'Sensuous solidarities: emotion, politics and performance in the Clandestine Insurgent Rebel Clown Army,' *Antipode* 44(2): 428–452.
Russell, B., Pusey, A. and Sealey-Huggins, L. (2012) 'Movements and moments for climate justice: from Copenhagen to Cancun via Cochabamba,' *ACME: An International E-Journal for Critical Geographies* 11(1): 15–32.

Russell, B., Schlembach, R. and Lear, B. (2017) 'Carry on camping? The British Camp for Climate Action as a political refrain,' in G. Brown, A. Feigenbaum, F. Frenzel and P. McCurdy (eds.), *Protest Camps in International Context: Spaces, Infrastructures and Media of Resistance*, Bristol: Policy Press, pp. 147–162.

Rutledge, I. and Wright, P. (1985) 'Coal worldwide: the international context of the British miners' strike,' *Cambridge Journal of Economics* 9(4): 303–326.

Sassen, S. (1991) *The Global City: New York, London, Tokyo*, Princeton, NJ: Princeton University Press.

Schneider, H. (2002) 'On the fault-line: the politics of AIDS policy in contemporary South Africa,' *African Studies* 61(1): 145–167.

Scholz, S.J. (2008) *Political Solidarity*, University Park: Pennsylvania State University Press.

Seekings, J. (1996) 'The 'lost generation': South Africa's youth problem in the early 1990s,' *Transformation* 29(1): 103–125.

Seekings, J. (2000) *The UDF: A history of the United Democratic Front in South Africa 1983–1991*, Oxford: James Currey.

Selbin, E. (2010) *Revolution, Rebellion, Resistance: The Power of Story*, London: Zed Books.

Sharma, S. (2014) *In the Meantime: Temporality and Cultural Politics*, Durham, NC: Duke University Press.

Sharp, J. (2013), 'Geopolitics at the margins? Reconsidering genealogies of critical geopolitics,' *Political Geography* 37: 20–29.

Shove, E., Pantzar, M. and Watson, M. (2012) *The Dynamics of Social Practice: Everyday Life and How It Changes*, London: Sage.

Shove, E., Trentmann, F. and Wilk, R. (2009) *Time, Consumption and Everyday Life: Practice, Materiality and Culture*, Oxford: Berg.

Shubin, V. (2008) *The Hot 'Cold war': The USSR in Southern Africa*, London: Pluto.

Simone, A. (2004) 'People as Infrastructure: Intersecting Fragments in Johannesburg,' *Public Culture* 16(3): 407–429.

Skelton, T. (2010) 'Taking young people as political actors seriously: opening the borders of political geography,' *Area* 42(2): 145–151.

Skelton, T. (2013a) 'Children, young people and politics: transformative possibilities for a discipline,' *Geoforum* 49: R4–R6.

Skelton, T. (2013b) 'Young people's urban im/mobilities: relationality and identity formation,' *Urban Studies* 50(3): 467–483.

Skinner, R. (2010) *The Foundations of Anti-Apartheid. Liberal Humanitarians and Transnational Activists in Britain and the United States, c. 1919 – 64*, Basingstoke: Palgrave Macmillan.

Smith, E. (2016) 'National liberation for whom? The postcolonial question, the Communist Party of Great Britain, and the Party's African and Caribbean Membership,' *International Review of Social History* 61: 283–315.

Smith, E. and Worley, M. (eds.), (2014) *Against the Grain: The British Far Left from 1956*, Manchester: Manchester University Press.

Southwood, I. (2011) *Non-Stop Inertia*, Winchester: Zero Books.

Stevens, S. (2014) 'Why South Africa? The politics of anti-apartheid activism in Britain in the long 1970s,' in J. Eckel and S. Moyn (eds.), *The Breakthrough. Human rights in the 1970s*, Philadelphia: University of Pennsylvania Press, pp. 204–225.

Stoller, P. (1997) *Sensuous Scholarship*, Philadelphia: University of Pennsylvania Press.

Suttner, R. (2008) *The ANC Underground in South Africa to 1976. A Social and Historical Study*, Auckland Park: Jacana.

Taft, J.K. (2011) *Rebel Girls: Youth Activism & Social Change across the Americas*, New York: New York University Press.

Thomas, S. (1996) *The Diplomacy of Liberation: The Foreign Relations of the ANC Since 1960*, London: IB Tauris

Thörn, H. (2006) *Anti-Apartheid and the Emergence of a Global Civil Society*, Basingstoke: Palgrave Macmillan.

Thörn, H. (2009) 'The meaning(s) of solidarity: narratives of anti-apartheid activism,' *Journal of Southern African Studies* 35(2): 417–436.

Tilly, C. (1999) 'Conclusion: from interactions to outcomes in social movements,' in M. Guigni, D. McAdam and C. Tilly (eds.), *How Social Movements Matter*, Minneapolis: University of Minnesota Press, pp. 253–270.

Tilly, C. and Tarrow, S. (2007) *Contentious Politics*, Oxford: Oxford University Press.

Toohey, P. (2011) *Boredom: A Lively History*, New Haven, CT: Yale University Press.

Trewhela, P. (1995) 'State espionage and the ANC London office,' *Searchlight South Africa* 3(4): 42–51.

Trewhela, P. (2009) *Inside Quatro: Uncovering the Exile History of the ANC and SWAPO*, Auckland Park: Jacana.

Tucker, A. (2016) 'Reconsidering relationships between homophobia, human rights and HIV/AIDS,' in G. Brown and K. Browne (eds.), *The Routledge Research Companion to Geographies of Sex and Sexualities*, London: Routledge, pp. 295–304.

Turner, A.W. (2010) *Rejoice! Rejoice! Britain in the 1980s*, London: Aurum.

Valentine, G. (2003) 'Boundary crossings: transitions from childhood to adulthood,' *Children's Geographies* 1(1): 37–52.

Valentine, G. (2008) 'Living with difference: reflections on geographies of encounter,' *Progress in Human Geography* 32(3): 323–337.

Vanderbeck, R.M. (2007) 'Intergenerational geographies: age relations, segregation and re-engagements,' *Geography Compass* 1(2): 200–221.

Van Kessel, I. (2000) *"Beyond Our Wildest Dreams" The United Democratic Front and the transformation of South Africa*, Charlottesville: The University Press of Virginia.

Vasudevan, A. (2017) *The Autonomous City. A History of Squatting*, London: Verso.

Virdee, S. (2014a) 'Challenging the empire,' *Ethnic and Racial Studies* 37(10): 1823–1829.

Virdee, S. (2014b) *Racism, Class, and the Racialized Outsider*, Basingstoke: Palgrave Macmillan.

Watt, P. (2016) 'A nomadic war machine in the metropolis: en/countering London's 21st-century housing crisis with Focus E15,' *City* 20(2): 297–320.

Weeks, J. (2000) *Making Sexual History*, Oxford: Wiley & Sons.

Weeks, J. (2007) *The World We Have Won: The Remaking of Erotic and Intimate Life*, London: Routledge.

Westad, O.A. (2005) *The Global Cold War: Third World Interventions and the Making of Our Times*, Cambridge: Cambridge University Press.

Wills, J., Datta, K., Evans, Y., Herbert, J., May, J. and McIlwaine, C. (2010) *Global Cities at Work: New Migrant Divisions of Labour*, London: Routledge.

Williams, E. (2012) 'Anti-apartheid: the Black British response,' *South African Historical Journal* 64(3): 685–706.

Williams, E. (2015) *The Politics of Race in Britain and South Africa: Black British Solidarity and the Anti-apartheid Struggle*, London: IB Tauris.

Wilson, H.F. (2012) 'Living with difference and the conditions for dialogue,' *Dialogues in Human Geography* 2(2): 225–227.

Wilson, H.F. (2016) 'On geography and encounter: bodies, borders and difference,' *Progress in Human Geography* online early. doi:10.1177/0309132516645958.

Wood, B.E. (2012), 'Crafted within liminal spaces: young people's everyday politics,' *Political Geography* 31: 337–346.

Wood, N., Duffy, M. and Smith, S.J. (2007) 'The art of doing (geographies of) music,' *Environment and Planning D: Society and Space* 25(5): 867–889.

Workers Aid for Bosnia (n.d.) *Taking Sides Against Ethnic Cleansing in Bosnia*, Leeds: Workers Aid for Bosnia.

Worth, N. (2009) 'Understanding youth transition as 'becoming': identity, time and futurity,' *Geoforum* 40(6): 1050–1060.

Wright, M.W. (2009) 'Justice and the geographies of moral protest: reflections from Mexico,' *Environment and Planning D: Society and Space* 27(2): 216–233.

Index

Abbott, Diane 158
Aebi, Nicole 150, 178, 180–1, 206
African National Congress (ANC) 3, 4, 5, 6, 10, 160, 186; banning of 4; City Group, relationship with 23, 24, 26, 39, 189; Defiance Campaign (1952) 3; unbanning of 176; Youth League 3
Afrikaner National Party 1
ANC *see* African National Congress (ANC)
Ansari, Selman 130, 205
Anti-Apartheid Movement (AAM) 23, 56; ANC support 6; Botha visit, protest of 26–7; City Group, conflict with 24, 26, 29–30; international 5–6, 10, 32; in South Africa 3–5; Wembley Stadium concert 159, 160; *see also* City Group
Anti-Fascist Action 88
apartheid: end of 32; legal structure for 2; opposition to 3–6; origins of 1–2; *see also* Anti-Apartheid Movement (AAM)
Armed Resistance Movement 4
arrests: legal support upon 121–3, 126, 131–4, 140, 143, 219; number of 105, 117, 120, 127–9, 134
Atkins, Sharon 106
Azanian People's Organisation (AZAPO) 82, 137

Balfe, Richard 27, 158
Banks, Lynne Reid 106
Banks, Tony 28, 118
banner 43, 48–50, 51
Barnett, Henry 23
Barnett, Mary 23

Batucada Mandela samba troupe 162, 196
Bearn, Mark 61
Benn, Tony 140, 177
Bickerstaffe, Rodney 27
Biko, Steve 4, 87, 162
Black Consciousness Movement (of Azania) (MCMA) 4, 6, 82, 160, 162, 185; City Group relationship with 39, 155
Black Londoners 92, 93
Bodden, Ken 198, 208; *see also* Hughes, Ken
Botha, President PW 7, 10, 26
Bowles, Adam 114, 115, 121–2, 215
boxes, storage 50
Boycott Movement 5
Breadline Cafe 55
Brickley, Carol 6–7, 23, 38, 44, 56, 73, 83, 103, 104, 110, 111, 117, 127, 154, 173, 174–5, 176, 179, 191, 196, 211
Brown, Gavin 16, 99n2
Brownlie, Fiona 65
Budd, Zola 135
Burgess, Mike 59–60, 66, 202, 212
Burma Campaign UK 208–9

Café in the Crypt 55
Caller, Hannah 208, 211
Camden New Journal 117
Canavan, Dennis 118
Cave, Eleanor 57, 112
Césaire, Aimé 4
Cherfas, Sidney 23
Chief Steward 73–5, 110, 129–30, 132, 156, 218
Chisholm, Sharon 57, 97, 132, 141, 148, 207
Christmas Day 88–90

City Group 15, 16; ANC, relationship with 23, 24, 26, 39, 189; Annual General Meeting 174–6; Anti-Apartheid Movement, relationship with 24, 26, 29–30; educational efforts by 84, 153–4; "Free Moses Mayekiso" campaign 86; "Free the Uppington 26" campaign 86, 163; fundraising concert 159–60; "Hands off Black Activists" 198; "Hands off Women Picketers" 134; 1982 picket 23–4, 25, 84; 1985 protest 31–2; non-sectarian approach 38, 39, 162, 184, 218; *Non-Stop Against Apartheid* 52, 137, 179–80; *Non-Stop News* 52, 134; "No Rights? No Flights!" 127, 135, 136, 137–41; *100 days and nights: a record of police harassment* 105; organisation of 73, 138; origins of 21, 23; *Picketers' News* 154; 'Picket University' 84; police relationship with 18, 109–12, 144; practices 25–32; *Record of Police Harassment on and around the NSP, 28th September 1989–4th January 1990* 129–31; solidarity within 34, 36, 37, 54, 64, 147; South African Embassy Picket Campaign (SAEPC) 28; sporting event protests 135, 141–3; Springbok 9 187–8; Springbok Reception Committee 184, 185–8; Surround the Embassy protest 161–3; trolley protests 135–7, 183; Uppington 14 campaign 183; user guide 73–4; weekly meeting 82–3; Youth and Students sub-group 154, 157, 159; *see also* Non-Stop Picket
City Group Singers 53–4, 154
City Limits 20
City of London Anti-Apartheid Group *see* City Group
Clark, Cate 28
Class War Federation 211
clipboards 51
Cohen, Harry 118
Collins, Amanda 73, 84, 116, 134
Commissioner's Directions 113, 118–20, 128
Committee of African Organisations 5
Communist Party of South Africa (CPSA) 3
Comprehensive Anti-Apartheid Act (US) 5

Congress Alliance 3
Congress of Democrats 3, 5, 22
Congress of South African Trade Unions (COSATU) 137
Control of Pollution Act 105
Corbyn, Jeremy 27, 28, 89, 158
Cowdrey, Chris 142–3
Crossman, Bob 89

Dark, Superintendent 27
de Klerk, President FW 174, 176, 186
Denver, Liz 214
Dismore, Andrew 106
donations 51–2, 106
Du Bois, WEB 3

Ebrahim, Gora 82
Elliot, Ann 25–6, 76, 77, 198–9, 201
embassy *see* South Africa House
End Loans to South Africa (ELTSA) 6
engagement with public 217
Evans, Ben 149–50, 151–2, 205

Fanon, Franz 4
Farmaner, Mark 51, 58, 62, 77, 81, 152, 183, 202, 206, 208–9, 214; *Record of Public Harassment* 129, 131
Featherstone, David 33, 35, 36, 37
Federation of South African Trade Unions (FOSATU) 9–10
Fernand, Kathy 63, 88, 138–9, 208
Fieldhouse, Roger 24, 30, 186
Forrest, Tunde 63, 199
Freedom Charter 3–4, 86–7
Freedom Songs 54
Free Steven Kitson campaign 23

Galbraith, Ross 186, 187, 191–2
Garavaldi, Betta 63, 178
Gargoyles, The 159
Garvey, Marcus 3
Gatting, Mike 141, 143
Giddens, Anthony 15
Gifford, Lord 25
Gilbertson, David 104, 110–11, 132
Gill, Ken 24
Godfrey, James 44, 122, 149, 155
Guardian 95
Gunnarsdóttir, Sigþrúður 62–3

Hain, Peter 27
Häkli, Jouni 14
Halt All Racist Tours (HART) 141

harassment 93, 103–9, 129–31, 198–9; anniversary celebration 90–1; homosexual 131; infrastructure restrictions 107–8; *100 days and nights: a record of police harassment* 105; racial 130–1, 198, 199; *Record of Police Harassment on and around the NSP, 28th September 1989–4th January 1990* 129–31; sexual 84, 134
Healy, Deirdre 47, 58, 77–8, 97, 166, 181, 203, 204, 205
Healy, Gerry 209
Higginbottom, Andy 31, 73, 89, 98, 106, 135, 208
Holland, Stuart 28
Hopkins, Peter 14
Horns of Jerico 84, 162
Houston, Whitney 159
Howlett, Commander 27, 28
Hughes, Ken 53–4; *see also* Bodden, Ken

Independent, The 187

Jewesbury, Daniel 60, 67, 165, 189, 210
John, Charine 155, 203, 206
Justice for Kitson Campaign 23, 24, 167

Kallio, Kirsi P. 14
Kay, Jackie 84
Kelliher, Diarmaid 36
Kempster, Jon 91–2, 93
Kenny, Dave 92
Kgokong, Mpotseng Jairus 185
Kitson, Amandla 21, 23
Kitson, David 21, 22, 28, 89, 92, 106, 154, 158, 192, 212; arrest of 120; incarceration of 22, 23–4; SACP relationship with 22, 24
Kitson, Major General Frank 110
Kitson, Norma 16, 22–3, 24, 25, 26, 73, 91, 106, 117, 119, 121, 127, 135, 154, 158, 162, 199, 215; ANC relationship with 21, 22, 23, 24, 26; arrest of 120; detention of 23; interview September 1986 1; launch of Non-Stop Picket 25–6, 43
Kitson, Steven 16, 23, 66, 90, 112, 117, 154; arrest of 28; detention of 21–2, 23
Konrad, Steve 139

Landau, Helen 188
Lansbury, Georgina 57, 66, 75, 97, 132, 141–2, 148, 164, 190

Lee, Police Inspector David 53
legal support 121–3, 126, 131–4, 140, 143, 219
Leigh, David 121
Liberal Party 4
Little, Superintendent 118–19, 120
Livingstone, Grace 58, 89, 133, 153, 156, 157, 158–9, 164, 165, 207, 213
location: importance of 216; use of 216; *see also* Trafalgar Square
London 10–13, 17; West End 54–5; youth in 13–15; *see also* Trafalgar Square
London School of Economics protest 158
Lowe, Gary 80, 150, 204

McHallem, Chris 159
Mail on Sunday's You 61
Maloney, Simone 76, 97, 134, 191
Mandela, Nelson 3, 24, 95–6; campaign to release 19; release of 19, 173, 176–7, 179; Wembley Stadium speech 184
Mandela, Winnie 179
Manley, Nick 80, 191, 205
March for Mandela 196
Marsden, Helen 51, 57–8, 79, 83, 88, 94, 96, 138, 159, 160, 168, 192, 206
Massey, Doreen 11, 34–5
May, Rufus 81
megaphone 52, 77, 104, 105, 106, 107
Meli, Francis 24
Mellon, Maggie 200–201
Menear, Inspector 28
Metropolitan Police 15, 50, 79–80; Albany Street Police Station 27; Cannon Row Police Station 27, 78, 84, 89, 110, 129, 134; Charing Cross Police Station 129; contention points 103–9; relations with City Group 18, 109–12, 144; Territorial Support Group 113, 163; *see also* harassment
Metropolitan Police Act (1839) 105
Michael, George 159
Minczer, Irene 66, 113–15, 116, 121, 156–7, 166, 167–8, 204
Minczer, Liz 113, 121
Miners' Strike of 1984–1985 57
Mitchell, John 162, 163
MK *see* Umkhonto we Sizwe (MK)
Mlambo, Johnson 82
Moodley, Strini 82
Morland, Judge 121
Morley's Hill 47, 51, 69, 216

Mothopeng, Zephania 93, 162
Motlatsi, James 135
Movement for Colonial Freedom 5
Murray, Simon 50, 60, 110, 139, 140, 168–9, 170, 182, 192–3
Myers, Bob 209
Myers, Liz 66
Mzontane, Rodwell 174

Namibia Support Committee 6
National Union of Students (NUS) 31
Neale, Graham 65
Netherlands, The 5
Nielsen, Rikke 150–1
Nkadimeng, Don 185
Nkosi Sikelel' iAfrika 144
noise 52–4; singing; protest of 104–7; *see also* megaphone
Non-European Unity Movement 4
Non-Stop Picket 1, 17–19, 41; participants in Non-Stop Picket; afterlife of 19, 196–219; anniversary celebrations on 90–4; annual commemorations 85, 86–90, 161–4; award 20–21; ban against 27–9, 116, 117–23, 128; creativity on 78; daily rhythms of 76–8; demands of 37–8; end of 19, 173–93; geographies of 47–8, 54–5; inclusiveness of 65, 151, 217; infrastructure of 48–51; intergenerational actions on 167–9; leaflet 42; lessons from 216–19; night shift on 78–81; noise of 52–4, 104–7; 1000 days celebration 94–6; organisation of 73–5, 217–18; origins 6–7, 10–11, 43–4, 196; paint throwing by 112–16, 121–3; practices of 37, 51–4, 72; rules of 73–4, 75; solidarity of 33, 35, 36; sustained commitment of 85–6, 174–6; temporalities of 72–99; unruly actions, reasons for 126–7; weekly rhythms of 81–5; women's picket 84; *see also* City Group
Noorani, Ruby 130, 200
"No Rights? No Flights!" protest 135, 136, 137–41

O'Donnell, Sally 59, 79, 127
Odeleye, Solomon 142

PAC *see* Pan-Africanist Congress (PAC)
Pan-Africanist Congress (PAC) 6, 82, 160; APLA 162; banning of 4; City Group relationship with 39; origins of 3, 4; unbanning of 176
Pandit, Shereen 210
participants in Non-Stop Picket 21, 55–61; levels of involvement in 65–8, 96–9; recruiting 52, 61–5, 154, 217; youthful 148–53; *see also* arrests
Patel, Haroon 82
Patel, Hema 96
Perry, Inspector 106
petition 51–2, 90, 94
Pheto, Molefe 92
Philo Chris 14
picket *see* Non-Stop Picket
Pityana, Barney 4
police *see* Metropolitan Police
Poll Tax campaign 210
practices 40; of City Group 25–32; definition of 40; elements of 40; Non-Stop Picket 37, 51–4, 72; reworking 212–16
Prevention of Terrorism Act 130
Prior, Arthur 23
Prior, Lena 23, 85
Privett, Andy 48, 106, 182, 187, 189–90, 210
Public Order Act 113, 117

Ramaphosa, Cyril 135
Rayne, Trevor 56, 76
RCG *see* Revolutionary Communist Group (RCG)
Reed, David 154; *see also* Yaffe, David
Reid, Lorna 56, 66, 76, 107, 113, 122–3, 133, 191–2, 207–8, 210, 212
remembrance rituals 82, 86–7
Remembrance Sunday 87–8
Revolutionary Communist Group (RCG) 56, 175, 200, 211; *Fight Racism! Fight Imperialism!* 30, 52; Rock Around the Blockade (RATB) 208
Revolutionary Communist Party 211–12
Reynolds, Penelope 59, 67, 79, 109, 117–18, 170, 205, 210
Richards, Chief Superintendent 28
Richardson, Jo 106
Roberts, Allan 118
Roberts, Ernie 27
Roques, Richard 29, 66, 74, 142, 181–2, 186, 191, 208
Rose, Gary 92

SACP *see* South African Communist Party (SACP)
St. Martin-in-the-Fields 113, 216; Café in the Crypt 55
Scholz, Sally J. 33–4
Schott, Andre 50, 75, 149, 152, 156, 164–5, 173, 181, 199, 203
Sedley, Stephen 28–9
Senzenina 82
Sharpeville Massacre 4, 5, 87
Sheriff, Gary 186
shopping trolley protests 135–7, 183
Shove, Elizabeth 37, 39–40, 41, 213
Simon Community 63–4
Simpson, Danny 130, 133, 143
singing 52–4, 78, 82, 144
Sisulu, Walter 3, 24, 176
Slovo, Joe 24
Smith, Fiona 14
Smith, Solly 24
Smith, Tony 159
Sobukwe, Robert Mangaliso 3
solidarity 147, 213; City Group, within 34, 36, 37, 54, 64, 147; creativity and 78; definition of 33; geographies of 32–6; growing up and 37; non-sectarian 38, 39, 162, 184, 218; as political struggle 36, 218–19; post picket 183–8; practising 36–43; singing as 54; youth, among 7, 147, 159–60
South Africa: brutality in 7; Constitution Act (1983) 8; Group Areas Act (1950) 2; homelands 2; Immorality Act (1950) 2; Population Registration Act (1950) 2; Promotion of Bantu Self-Government Act (1959) 2; Reservation of Separate Amenities Act (1953) 2; Sharpeville Massacre 4, 5; Soweto Uprising 4–5; township uprisings 9, 10; Tricameral Parliament 7, 9; *see also* apartheid; South Africa House
South Africa House 18, 20, 38, 47; chaining to 127, 163; closure of demand 37, 38; flowers left at 82, 163; opposition to protest 104–5, 108–9; paint throwing on 27, 112; protest of banned 27–9, 116, 117–21, 128; scrubbing 127
South African Airways (SAA) "No Rights? No Flights!" protest 127, 135, 136, 137–41

South African Communist Party (SACP) 4, 22, 24; unbanning of 176
South African Council on Sport (SACOS) 141, 185
South African Indian Congress 3, 5
South African Native National Congress (SANNC) 3
South African Youth Congress 137
South Africa – the Imprisoned Society (SATIS) 6
Soweto Uprising 4–5, 87; commemoration of 161–4
sporting events, protests of 135, 141–3, 184, 185–8
Springbok 9 187–8
Springbok Reception Committee 184, 185–8
Squire, Francis 49, 50, 55, 64–5, 83, 168, 177, 181, 187
States, Adrian 117
Stop the Seventy Tour 141
storage boxes 50
supermarket trolley protests 135–7, 183
Sutton, Jacky 62, 78–9, 178, 204

Tambo, Oliver 3
Tatchell, Peter 106, 162
Terry, Mike 28
Thackray, Dominic 95, 111, 131, 142, 143, 158, 159–60
Thörn, Hakan 5, 32
Tilly, Charles 41
Trafalgar Square 20, 37, 43, 47, 49, 69, 78–81, 182, 216

UDF *see* United Democratic Front (UDF)
UK Uncut 210
Umkhonto we Sizwe (MK) 4, 22
Unite Against Fascism 210
United Democratic Front (UDF) 7; banning of 137; origins of 7–10
United States 5
unruliness 18, 126–44; empowerment through 126, 143–4; reasons for 126–7

van Kessel, Ineke 8, 9
Vienna Convention 28, 108

Waller, Rene 23, 95, 97, 168
Wedding Present, The 159, 160
Weinberg, Joan 23

Weiner, Cat 67, 134, 139, 140, 187–8, 200, 204–5, 208, 211, 214–15
Westminster noise pollution bylaws 105, 107
Workers Aid for Bosnia 209–10

Yaffe, David 43, 56, 83, 119, 215; *see also* Reed, David
Yaffe, Helen 16
Yaffe, Susan 56, 163, 169, 190
youth 57, 169–71; agency of 153–4; British 11, 13–15; demographics of 148–53; empowerment of 14, 126, 151; growing up 15, 18, 37, 41, 164–6; intergenerational interactions 167–9; levels of participation by 96–9; organising 157–9; responsible tasks for 155–7; solidarity amongst 7, 147, 159–60; in South Africa 7, 8–9; vulnerability of 197–202
Youth Campaign for Nuclear Disarmament (YCND) 57

Zolile Hamilton Keke 82, 89, 121, 162